Topics In Mathematical Physics

SKY AND WATER I. Woodcut by M.C. Escher. Courtesy Vorpad Galleries, San Francisco, Chicago, New York, and Laguna Beach

TOPICS IN MATHEMATICAL PHYSICS

Papers presented at an International Symposium held July 28-August 2, 1975, at Boğaziçi University, Istanbul, Turkey

Edited by
HALİS ODABAŞI and Ö. AKYÜZ
Boğaziçi University, Istanbul

COLORADO ASSOCIATED UNIVERSITY PRESS
BOULDER, COLORADO 1977

1424 Fifteenth Street
Boulder, Colorado 80309
ISBN 0-87081-072-3
Library of Congress Card Catalog 77-84853
Printed in United States of America

CONTENTS

PREFACE

In order to survey the recent progress in mathematical physics, an International Symposium was held on the campus of Bogazici University, Istanbul, Turkey, during the week of July 28-August 2, 1975. The Symposium was mainly financed by Bogazici University, and a small grant was also received from the Scientific and Technical Research Council of Turkey. Since it was impossible to cover the entire field of mathematical physics in such a short time, we wanted to choose certain topics in which, we thought, there had been exciting and rapid progress in recent years. So, Continuum Mechanics, Group Representations, Statistical Mechanics and Field Theories were chosen. This book contains the talks given by the invited lecturers on these topics.

Perhaps it should be explained why Continuum Mechanics was chosen as a topic in mathematical physics. It is a well-known fact that in recent years the methods and concepts of Continuum Mechanics have been finding more and more application in several branches of mathematical physics. Therefore, it was thought that the inclusion of the topic would be very beneficial to the interested physicists.

We would like to take this opportunity to thank Prof. A. Kuran, President of Bogazici University, for his generous financial support which made the Symposium and this book possible. We would also like to thank Prof. R. Berker, the chairman of the organizing committee, for his invaluable help and encouragement. Thanks are also due to Miss Ş. Türsan for her tireless efforts in the organizing of the Symposium and later in retyping the manuscripts uniformly. Last, but not least, we thank each one of the lecturers for their participation and for their contributions of which this book is made.

Halis Odabaşi

R. Ömür Akyüz

OPENING ADDRESS

APTULLAH KURAN, PRESIDENT

BOGAZICI UNIVERSITY

Distinguished Lecturers and Guests, Ladies and Gentlemen;

It is a singularly happy occasion for me to welcome all of you to the International Symposium on Mathematical Physics. This Symposium will be the first of a series of Summer Symposia that will be held each year on this campus. It is most fitting that we start the series with a scientific subject; for pure science has been one of the major areas of development for Bogazici University. Despite continuing student disinterest in science as a field of study, an attitude that I hope will soon change, our policy has been one of expansion. Since 1971 the number of teaching staff in our departments of mathematics, physics, and chemistry have doubled. Working drawings for a new laboratory complex are ready and construction will begin in early Fall.

Laboratory equipment for the new facilities are already begin purchased, and, if all goes well, the new building will be in full operation in three years time. A serious gap in our science curriculum will soon be remedied, for during the past academic year it was decided by the University Executive Committee to establish two new departments - those of Biology and Computer Science. Both will be activated at the beginning of the 1976-77 academic year.

One of the main functions of a university is obviously teaching, which is carried on largely by the teaching staff of that institution. But a second function of equal importance is to foster research, which by its very nature, transcends the boundaries of a particular university. Communication is the only tool for exchange of ideas; the most efficient form of communication being direct person to person contact. If we are to find solutions to the numerous problems facing us, scholars and scientists have to be in touch with one another and meet as often as they can. Scholarship and science know no geo-political frontiers. They are, and have always been the foremost agents of international cooperation and understanding. It gives me great pleasure, therefore, to see many scientists from various countries who are gathered here today to associate with each other, to assist each other and to benefit from each other. I am confident that this collaboration will be very fruitful for everybody involved. I am also confident that the lecturers and participants of the Bogazici University Summer Symposia will make a contribution to science and scholarship this year and in the years to come.

Before I conclude, I would like to express my heartfelt thanks to the organizing committee whose members worked long and hard to implement the idea that was put forth, as well as all the persons who generously donated their time and effort to make your stay here a pleasant and memorable one. I would also like to thank the lecturers and the participants of this Symposium for kindly accepting our invitation and wish all of you a productive and enjoyable week at Boğaziçi University.

FUNDAMENTALS OF CONTINUUM FIELD THEORIES

A. Cemal Eringen

Princeton University
Princeton, New Jersey

ABSTRACT

Mathematical methods for constructing classical and contemporary continuum theories are presented and future trends are indicated.

1. INTRODUCTION

All branches of physical sciences are based on one or the other of the two following fundamental models: (a) Atomic (or discrete); (b) Continuum (or field). The first model considers that all bodies are made up of discrete points which are endowed with some basic physical notions (e.g. mass and charge) and the second assumes that the geometrical points of bodies are endowed with continuous fields. Both models have registered great successes in their domains of applications, beyond the imagination of its originators. Nevertheless, each model alone fails to explain large classes of important physical phenomena that fall between the two extremes, the atom and the continuum. While clearly connections and bridges must exist for passage from one to the other uncovering and explaining a great many interesting problems in the intermediary regions, these bridges have not been constructed yet, except in special and ideal situations. Mathematical difficulties for the application of lattice dynamics and quantum theories (that constitute the basis of the discrete model) to real bodies are so great and the cost is so high that we cannot hope to carry out calculations for real materials. On the other hand, classical continuum field theories fail to apply in the microscopic and atomic ranges. Thus, for example, for materials with inner structure (e.g., composites, liquid crystals, granular and porous materials) new field theories are needed in order to account for their extra degrees of freedom over those present in the classical theories.

The domain of application of any physical theory may be determined by the ratio of two characteristic lengths (or characteristic times) one associated with the external effects, L, (e.g. wave length) and the other with the inner structure, ℓ, (e.g. average granular or atomic distance). When $\ell/L \ll 1$ then the material points of the body act collaboratively so that statistical averages are important and the body behaves as a continuum. However, for $\ell/L \sim 1$ we are forced to consider the identity of material points (or subbodies) in a body. This is the case certainly for *all* bodies in any case, fortunately however, such complications are not always warranted in a given problem. Nevertheless, in many problems, the effect of the inner structure of bodies presents itself in a dramatic fashion (e.g. fracture, surface physics). Thus sooner or later, we must incorporate the effects of the inner structure of the body and face the major task of unifying the continuum and atomic theories.

In accordance with the suggestion of our gracious host Professor R. Berker, in the first of these three lectures I shall discuss briefly the mathematical formulation of classical field theories. The second lecture is devoted to a brief exposition of polar field theories (relevant to bodies with inner structure) and the third to an aspect of the nonlocal field theories which are presently on the research stage. I shall try to demonstrate, with some critical examples, the success of both the polar and nonlocal field theories.

Unfortunately, physics and mathematics education in the U.S. during the past fifty years has left continuum theories an orphan. The result is that present material theories, which are good for ideal solids (e.g. periodic lattices), lack much that field theories can supply for the understanding of physics of real materials. Especially the last decade has provided such order and scope in fundamental ideas underlying continuum field theories, that other branches of physics (e.g. thermodynamics of irreversible processes, electromagnetics of deformable bodies, relativistic continua, micromagnetism) may benefit towards resolving many of the outstanding difficulties left to sleep at the turn of the century and open new inroads for significant advances. Moreover, with the opening of the new field of nonlocal continuum physics, the dream of constructing the impossible bridge connecting classical field theories to atomic physics is finally in sight. I am therefore pleased to have this opportunity to speak to this distinguished group of physicists, mathematicians, and engineers on the topic.

2. SUCCESS AND FAILURE OF CLASSICAL FIELD THEORIES

Three well-founded and widely used classical field theories are: Elasticity, Fluid Dynamics, and Electromagnetic Theory.

(i) *Theory of Elasticity*. The field equations of the linear theory of elasticity for the isotropic solids are the celebrated Naviers equations whose solutions give the displacement field $\mathbf{u}(\mathbf{x},\mathbf{t})$ of each point $\mathbf{x}$ of the body at time t:

$$(\lambda + 2\mu)\, \nabla\nabla \cdot \mathbf{u} - \mu\nabla \times \nabla \times \mathbf{u} + \rho(\mathbf{f} - \ddot{\mathbf{u}}) = \mathbf{0} \tag{2.1}$$

where λ and μ are the Lame constants, ρ the mass density and $\mathbf{f}$ is the body force density. The success of this theory may be summarized briefly as:

a. By means of (2.1), we have been able to determine the displacement and stress fields in elastic solids when they are subjected to external loads, boundary displacements or body forces.

b. Theories beams, plates, and shells have been developed and became independent fields by themselves. By means of these theories, it has been possible to calculate and design aerospace, mechanical and civil engineering structures, (e.g. airplanes, space vehicles, turbines, buildings, bridges).

c. Fracture mechanics is based on (2.1).

d. The entire field of geophysics, oil exploration makes use of (2.1) of its variants and modifications.

(ii) *Theory of Viscous Fluids*. The field equations of viscous fluids are the Navier-Stokes equations whose solution gives the velocity field $\mathbf{v}(\mathbf{x},\mathbf{t})$:

$$-\nabla p + (\lambda_v + 2\mu_v)\, \nabla\nabla \cdot \mathbf{v} - \mu_v\, \nabla \times \nabla \times \mathbf{v} + \rho(\mathbf{f} - \dot{\mathbf{v}}) = \mathbf{0} \tag{2.2}$$

where λ_v and μ_v are the viscosity coefficients and $p(\underset{\sim}{x},t)$ is the pressure. On the success of the Navier-Stokes theory, one may enumerate

a. The field of aerodynamics is based on (2.2).

b. The fields of Hydrodynamics, Hydrolics, Lubrication are all based on (2.2).

c. Atmospheric physics and oceanography make heavy uses of the Navier-Stokes theory.

d. Even suspensions, blood flow, and biodynamics lean heavily on (2.2).

(iii) *Electromagnetic Theory*. The field equations are the well-known Maxwell's equations whose solutions determine the electric and magnetic fields $\underset{\sim}{E}$ and $\underset{\sim}{H}$:

$$-\underset{\sim}{\nabla} \times \underset{\sim}{E} = \frac{1}{c}\dot{\underset{\sim}{B}} \quad , \quad \underset{\sim}{\nabla} \times \underset{\sim}{H} = \frac{1}{c}\dot{\underset{\sim}{D}} + \frac{4\pi}{c}\underset{\sim}{J}$$
$$\underset{\sim}{\nabla} \cdot \underset{\sim}{B} = 0 \quad , \quad \underset{\sim}{\nabla} \cdot \underset{\sim}{D} = 4\pi\rho_e \tag{2.3}$$

supplemented with constitutive equations (for isotropic media).

$$\underset{\sim}{D} = \varepsilon \underset{\sim}{E} \quad , \quad \underset{\sim}{B} = \kappa \underset{\sim}{H} \quad , \quad \underset{\sim}{J} = \sigma \underset{\sim}{E} \tag{2.4}$$

Here, c, ε, κ, σ and ρ_e are respectively the speed of light, the dielectric constant, permeability, electric conductivity and the charge density. From (2.3) and (2.4) by elimination of $\underset{\sim}{H}$ (for constant ε, κ, σ) we obtain the field equations for $\underset{\sim}{E}$, for example,

$$\underset{\sim}{\nabla} \times \underset{\sim}{\nabla} \times \underset{\sim}{E} + \varepsilon\kappa \ddot{\underset{\sim}{E}} + \sigma\kappa \dot{\underset{\sim}{E}} = \underset{\sim}{0} \tag{2.5}$$

Among many successful applications of the Maxwell E-M theory, we cite:

a. Propagations of radio waves, antenna theory

b. Optics

c. Magnetism

d. Electric circuits, computers

Indeed, today almost all fields of engineering and physics make use of these field theories one way or another and yet there exist many crucial problems for which these field theories fail to predict the observed phenomena. To cite a few: The problems of fatigue and fracture, so essential to designs with solids, remain unresolved. While we have some semi-empirical approaches to these fields the predictions of state of stress near the crack tip and the growth and accumulation of microcracks leaves much to be desired.

In the field of fluid dynamics, in spite of feverish activities, we have no rational approach to turbulence. Navier-Stokes theory is useless when we deal with polymeric substances and liquid crystals. Maxwell E-M theory falls short in the high frequency region. Dispersion, absorption and pletora of phenomena arising from high electric or magnetic fields (e.g. superconductivity, space charge distribution, electrets) cannot be dealt with.

In summary, classical field theories fail to apply for

A) Bodies subject to large fields (e.g. large deformation, high electric fields).

B) Memory dependent materials (e.g. polymers, viscoelastic materials).

C) Materially non-uniform bodies (e.g. inhomogeneous and anisotropic bodies, composites, bodies with inner structure).

D) Bodies subject to long range fields.

During the past 25 years, major progress has been made in the categories (A), (B), and (C). It can be said that the formulation is complete (more or less) for nonlinear field theories (A) and constitutive theories for memory dependent materials (B), although applications of these theories to physical problems and experimental work are meager. The classical approach to materially non-uniform bodies was through the concept of inhomogeneity and anisotropy which required introduction of tensor material moduli that depends on the position. This phenomenological approach while useful needed much deeper questioning. For example, it is well known (as a theorem) that Navier-Stokes fluids are *isotropic fluids*, and the tensor viscosity coefficients that can be introduced (just as in anisotropic elasticity theory) must be *isotropic tensors*. Hence, there cannot be more than two viscosity moduli. This means that contrary to the elasticity theory, there is no way we can modify the constitutive equations of Navier-Stokes fluids to account for the anisotropy. Yet experimental observations have shown that a large class of fluids is anisotropic (e.g. Liquid crystals, polymeric fluids).

These and similar needs for solids (e.g. structured solids, granular, porous, and fibrous materials, magnetic solids) provided stimulus for the development of theories of polar continua. By means of these theories, it is now possible to predict physical phenomena in the microscopic range. However, even the polar theories are not adequate for the understanding of many physical phenomena (e.g. crack tip stress, surface physics, high frequency waves, turbulence). During the last few years a limited amount of research work in continuum physics has been directed to understand and control the class of physical phenomena whose origin may be traced to the long range interatomic interactions. Already several reasonably satisfactory continuum formulations and some solutions exist indicating the power of the theory.

Clearly, it is impossible even to sketch the basic developments in any one of these topics, within the limited time allowable. Following the suggestion of my host, I shall devote two of my lectures to a brief review of the fundamental concepts and methodology in the contemporary nonlinear field theories and in my third lecture, I plan to present some of our recent work on the nonlocal continuum field theories.

For efficiency and continuity, mostly, I shall follow the theme of elastic solids through all four stages (A) to (D) with some remarks regarding fluids and memory dependent materials. For the growing fields of electromagnetism and multiple interactions of many fields that occur in mixture and E-M elastic solids and fluids, we refer the reader to Eringen [1976] and references therein. I shall not touch to relativistic field theories, although the literature here is abundant and impressive.

3. CHARACTERIZATION OF CONTINUA

For the establishment of the mathematical formulation of any physical phenomenon that takes place in a material body, it is necessary to identify the elements of a body together with their preassigned primitive characteristics and the operational rules that they are subject to. The body is identified as a collection of material points which are endowed with *mass* and *charge*. The operational rules are the *laws* of physics that are relevant to mass and charge. Specifically, the body is considered to occupy a spatial region B of the Euclidian space E_3 whose points may be referred to a set of cartesian coordinates X^K, $K = 1, 2, 3$ and a material point P at the *reference state* at time $t = 0$ may be identified with $X^K = X^K(P)$ so that no distinction need be made between P and X^K. The motion carries various points P in B to spatial places x^k at time t, Fig. 1. Thus *the motion is defined as a one parameter family of transformations.*

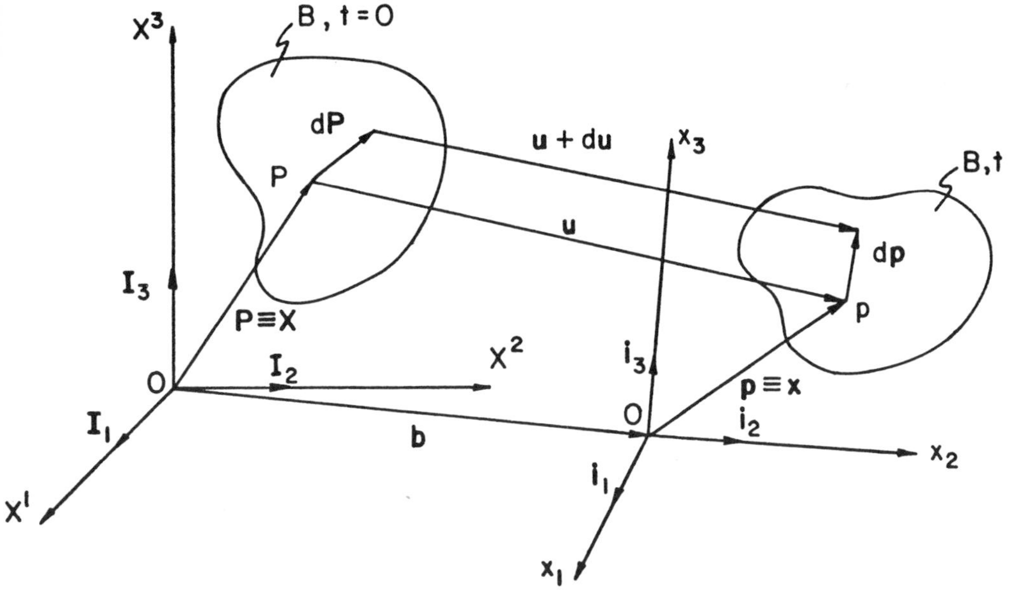

Fig. 1 Reference and Deformed States

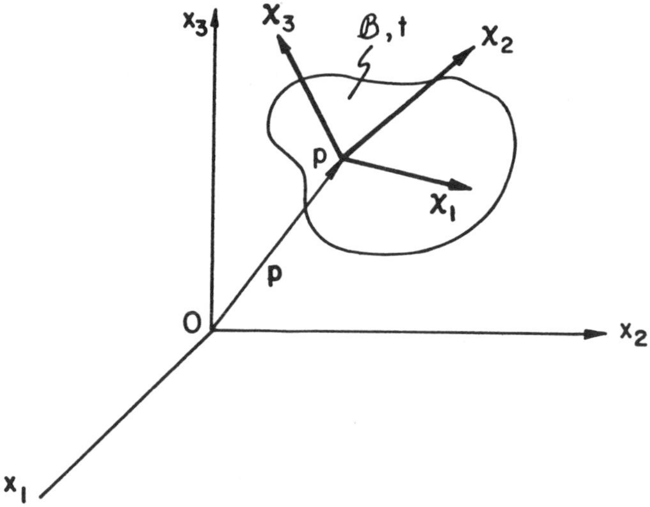

Fig. 2 Directors $\underset{\sim}{X}_K$ of Polar Continua

$$x^k = \hat{x}^k(X^K, t) \quad , \quad x^k \in \mathcal{B} \quad , \quad k = 1, 2, 3 \tag{3.1}$$

The *axiom of continuity* then states that the inverse (3.1) exists at all points x^k at all times, i.e.

$$X^K = X^K(x^k, t) \quad , \quad K = 1, 2, 3 \tag{3.2}$$

This is fulfilled if (3.1) possesses continuous partial derivatives and the jacobian

$$J \equiv \det (\partial x^k / \partial X^K) > 0 \tag{3.3}$$

at all points of the body, at all times. Physically the axiom of continuity implies that no region of body with finite volume can become one with zero or infinite volume. A finite material volume goes into a finite volume, a surface to a surface and a line to a line. This is the picture in all classical continuum theories where the material points are geometrical points endowed with physical notions of mass and charge.

It may happen that certain internal constraints exist among the material points, e.g. pairs of points may be rigidly attached to form dipolar elements. While such cases can be built up by imposing additional conditions on the motions of the individual material points of the pairs it is more convenient and much simpler to consider such elements as material points with extra degrees of freedom. For example, for bodies consisting of dipolar elements we may characterize an element of the body with a position vector $\underset{\sim}{X}$ in addition with three rigidly attached vectors ($\underset{\sim}{\chi}_K = \chi_K^k \underset{\sim}{i}_k$, K = 1, 2, 3, $\underset{\sim}{i}_k$ = unit base vectors) representing rotatory degrees of freedom ,Fig. 2. In this way we can disregard the conditions arising from the internal constraints. This is the case for *micropolar bodies*. Thus the primitive elements of the micropolar bodies are material points that possess rotatory degrees of freedom, hence they may be envisioned as small rigid bodies. In this case then the motion is fully characterized by (3.1) and

$$\chi_K = \hat{\chi}_K(X, t) \tag{3.4}$$

subject to

$$\chi \chi^T = I \quad ; \quad \chi \equiv (\chi^k{}_K) \tag{3.5}$$

where χ^T is the transpose of the matrix χ and I is the unit matrix. A material point in body now possesses three translational degrees of freedom (3.1) and three rotational degrees of freedom (3.4).

More generally we may lift the conditions (3.5). In this case, the dipolar elements are deformable and we have a deformable point with 12 degrees of freedom, three translations x^k, and nine microdeformation degrees of freedom characterized by $\chi^k{}_K$, without the constraints (3.5). This is the basis of the theory *micromorphic continua*. This scheme can be continued further to build up polar theories with higher degrees of freedom. However, soon this process becomes too tedious for any practical use. For further and deeper penetration to the atomic world it is necessary to resort to the nonlocal effects. Nevertheless, even in the atomic scale we may have to deal with polar molecules which require the use of the above concepts.

For the electromagnetic effects, beyond the classical concepts, it is necessary to make provision by introducing new vectors and tensor fields to characterize fields embodied in polar and deformable molecules. Similar to micropolar and micromorphic mechanics there exist polar E-M theories (cf. Eringen and Kafadar [1970], Maugin and Eringen [1972a,b].

4. STRAIN AND ROTATION MEASURES

The squares of arc lengths, $(ds)^2$ and $(dS)^2$ in the spatial and reference configurations $\mathcal{B}$ and B are calculated, respectively by

$$(ds)^2 = \delta_{k\ell}\, dx^k\, dx^\ell \quad , \quad (dS)^2 = \delta_{KL}\, dX^K\, dX^L \tag{4.1}$$

The motion (3.1) and its inverse (3.2) provide relations between the spatial and material line elements $d\underset{\sim}{p}$ and $d\underset{\sim}{P}$ (Fig. 1)

$$d\underset{\sim}{p} = \underset{\sim}{p}_{,K}\, dX^K = \underset{\sim}{C}_K\, dX^K \quad , \quad d\underset{\sim}{P} = \underset{\sim}{P}_{,k}\, dx^k = \underset{\sim}{c}_k\, dx^k \tag{4.2}$$

where

$$\underset{\sim}{C}_K(\underset{\sim}{X},t) \equiv \underset{\sim}{p}_{,K} = \underset{\sim}{P}_{,k}\, x^k_{,K} = \underset{\sim}{i}_k x^k_{,K} \;,\; \underset{\sim}{c}_k(\underset{\sim}{x},t) = \underset{\sim}{P}_{,k} = \underset{\sim}{P}_{,K}\, X^K_{,k} = \underset{\sim}{I}_K\, X^K_{,k} \tag{4.3}$$

where $\underset{\sim}{i}_k$ and $\underset{\sim}{I}_K$ are the unit cartesian base vectors in x^k and X^K. Substituting (4.2) into (4.1) we obtain

$$(ds)^2 = C_{KL}\, dX^K\, dX^L \quad , \quad (dS)^2 = c_{k\ell} dx^k\, dx^\ell \tag{4.4}$$

where the *symmetric positive definite* tensors

$$\begin{aligned} C_{KL}(\underset{\sim}{X},t) &\equiv \underset{\sim}{C}_K \cdot \underset{\sim}{C}_L = \delta_{k\ell}\, x^k_{,K}\, x^\ell_{,L} \\ c_{k\ell}(\underset{\sim}{x},t) &\equiv \underset{\sim}{c}_k \cdot \underset{\sim}{c}_\ell = \delta_{KL}\, X^K_{,k}\, X^L_{,\ell} \end{aligned} \tag{4.5}$$

are called *Green* and *Cauchy deformation tensors*, respectively. It is clear that the motion is rigid (with no deformation) for *arbitrary directions* if and only if

$$C_{KL} = \delta_{KL} \quad \text{or} \quad c_{k\ell} = \delta_{k\ell}$$

Therefore, $\underset{\sim}{C}$ and $\underset{\sim}{c}$ represent the measure of the local deformation in a neighborhood of the points P and p. The inverse matrices of $\underset{\sim}{C}$ and $\underset{\sim}{c}$ can also be used as the measure of deformation. These are given by

$$\overset{-1}{C}{}^{KL} \equiv B^{KL} = \delta^{k\ell} X^{K}_{,k} X^{L}_{,\ell} \quad , \quad \overset{-1}{c}{}^{k\ell} \equiv b^{k\ell} = \delta^{KL} x^{k}_{,K} x^{\ell}_{,L} \tag{4.6}$$

and are, respectively, called *Piola* and *Finger deformation tensors*.

Lagrangian and *Eulerian strain* tensors E_{KL} and $e_{k\ell}$ are defined by

$$2E_{KL} \equiv C_{KL} - \delta_{KL} \quad , \quad 2e_{k\ell} \equiv \delta_{k\ell} - e_{k\ell} \tag{4.7}$$

They are used to calculate change in the arc length due to deformation

$$(ds)^2 - (dS)^2 = 2E_{KL}\, dX^K\, dX^L = 2e_{k\ell}\, dx^k\, dx^\ell$$

and hence they are related to each other by

$$E_{KL} = e_{k\ell}\, x^{k}_{,K}\, x^{\ell}_{,L} \quad , \quad e_{k\ell} = E_{KL}\, X^{K}_{,k}\, X^{L}_{,\ell} \tag{4.8}$$

Introducing the displacement vector $\underset{\sim}{u}$ by (Fig. 1)

$$\underset{\sim}{u} = \underset{\sim}{p} - \underset{\sim}{P} + \underset{\sim}{b} \tag{4.9}$$

we can calculate

$$\begin{aligned} \underset{\sim}{C}_K &= \underset{\sim}{P}_{,K} + \underset{\sim}{u}_{,K} = \underset{\sim}{I}_K + U_{M,K}\, \underset{\sim}{I}^M \\ \underset{\sim}{c}_k &= \underset{\sim}{p}_{,k} - \underset{\sim}{u}_{,k} = \underset{\sim}{i}_k - u_{m,k}\, \underset{\sim}{i}^m \end{aligned} \tag{4.10}$$

Substituting these into (4.5) and (4.6) we obtain

$$\begin{aligned} C_{KL} &= \delta_{KL} + 2E_{KL} = \delta_{KL} + U_{K,L} + U_{L,K} + U^{M}_{,K}\, U_{M,L} \\ c_{k\ell} &= \delta_{k\ell} - 2e_{k\ell} = \delta_{k\ell} - u_{k,\ell} - u_{\ell,k} + u^{m}_{,k}\, u_{m,\ell} \end{aligned} \tag{4.11}$$

which give the exact *deformation* and *strain measures* (Note the nonlinear terms).

An alternative approach to strain measures is through the *polar decomposition theorem*, which is revealing. In matrix notation, any invertible linear transformation $\underset{\sim}{F}$ has two multiplicative decomponsitions

$$\underset{\sim}{F} = \underset{\sim}{R}\,\underset{\sim}{U} = \underset{\sim}{V}\,\underset{\sim}{R} \tag{4.12}$$

where $\underset{\sim}{R}$ is *orthogonal* and $\underset{\sim}{U}$ and $\underset{\sim}{V}$ are *symmetric positive matrices*. The following relations hold

$$\underset{\sim}{U}^2 = \underset{\sim}{F}^T\,\underset{\sim}{F} \quad , \qquad \underset{\sim}{V}^2 = \underset{\sim}{F}\,\underset{\sim}{F}^T$$
$$\underset{\sim}{V} = \underset{\sim}{R}\,\underset{\sim}{U}\,\underset{\sim}{R}^T \quad , \qquad \underset{\sim}{V}^2 = \underset{\sim}{R}\,\underset{\sim}{U}^2\,\underset{\sim}{R}^T \tag{4.13}$$

when $\underset{\sim}{F}$ is identified by the deformation gradient $x^k_{,K}$ then $\underset{\sim}{R}$ is the *rotation tensor* and $\underset{\sim}{U}$ and $\underset{\sim}{V}$ give the deformation measures, i.e.,

$$R^k{}_K = x^k{}_{,L}\,\overset{-\frac{1}{2}}{C}{}^L{}_K \quad , \qquad (\underset{\sim}{U}^2)^K{}_L = \overset{-1}{C}{}^K{}_L \quad , \qquad (\underset{\sim}{V}^2)^k{}_\ell = \overset{-1}{c}{}^k{}_\ell = b^k{}_\ell \tag{4.14}$$

Indeed $R^k{}_K$ can be shown to represent the finite rotation at a point. For a detailed discussion see Eringen [1962, 1967, 1975], Truesdell and Noll [1965].

For the discussion of fluids it is necessary to introduce *relative strain* and *rotation* measures. In this case the relative motion, $\underset{\sim}{\xi}$, of a material point $\underset{\sim}{X}$ at time τ with respect to the configuration at time $t \leq \tau$ is necessary

$$\underset{\sim}{\xi} = \underset{\sim}{x}(\underset{\sim}{X},\tau) \tag{4.15}$$

Introducing (3.2) for $\underset{\sim}{X}$, this gives

$$\underset{\sim}{\xi} = \underset{\sim}{x}(\underset{\sim}{X}(\underset{\sim}{x},t),\tau) \equiv \underset{\sim}{x}_{(t)}(\underset{\sim}{x},\tau) \tag{4.16}$$

where $\underset{\sim}{x}_{(t)}$ is called the *relative deformation function.* If we write $\underset{\sim}{F}_{(t)}(\tau)$ for $\partial\xi^{\alpha}/\partial x^{k}$ and $\underset{\sim}{F}^{(t)}_{(\tau)}$ for $x^{k}_{,\alpha}$ then the relative strains measures are

$$\underset{\sim}{c}_{(t)}(\tau) \equiv \underset{\sim}{F}_{(t)}(\tau)^{T}\, \underset{\sim}{F}_{(t)}(\tau) \quad , \quad b_{(t)}(\tau) \equiv \underset{\sim}{F}_{(\tau)}(t)\, \underset{\sim}{F}_{(\tau)}(t)^{T} \tag{4.17}$$

Strain and rotation measures for the polar bodies can be introduced similarly (cf. Eringen and Şuhubi [1964], Eringen [1966], Eringen and Kafadar [1975]).

5. KINEMATICS OF CONTINUA

Crucially important in the kinematics of continua is the time rates of fields with the material point $\underset{\sim}{X}$ held constant. This is called *material derivative.* Thus, for example, the *velocity,* $\underset{\sim}{v}$, and *acceleration,* $\underset{\sim}{a}$, of a material point $\underset{\sim}{X}$ are defined by

$$\underset{\sim}{v} = \frac{\partial \underset{\sim}{x}(\underset{\sim}{X},t)}{\partial t} \equiv \dot{\underset{\sim}{x}} \quad , \quad \underset{\sim}{a} \equiv \frac{\partial \underset{\sim}{v}(\underset{\sim}{X},t)}{\partial t} \equiv \dot{\underset{\sim}{v}} \tag{5.1}$$

through (3.2) we may write

$$\underset{\sim}{v}(\underset{\sim}{X},t) = \underset{\sim}{v}(\underset{\sim}{X}(\underset{\sim}{x},t),t) = \hat{v}(\underset{\sim}{x},t)$$

and therefore we can also write for

$$\underset{\sim}{a} = \dot{\underset{\sim}{v}} = \frac{\partial \hat{\underset{\sim}{v}}}{\partial t} + \hat{\underset{\sim}{v}}_{,k}\, \dot{x}^{k} \qquad \text{or} \qquad a^{k} = \left.\frac{\partial v^{k}}{\partial t}\right|_{\underset{\sim}{x}} + v^{k}_{,\ell}\, v^{\ell} \equiv \frac{Dv^{k}}{Dt} \tag{5.2}$$

This is the well-known *eulerian view-point* where the identity of the material point is lost and we have a field $a^{k}(\underset{\sim}{x},t)$ associated with each spatial point $\underset{\sim}{x}$. Similarly we can calculate material rates of other vector and tensor fields. Often useful in these calculations is

Fundamental Lemma. The material time rate of deformation gradients are given by

$$\frac{D}{Dt}(x^{k}_{,K}) = \dot{x}^{k}_{,K} = v^{k}_{,\ell}\, x^{\ell}_{,K} \qquad \text{or} \qquad \dot{\overline{dx^{k}}} = v^{k}_{,\ell}\, dx^{\ell} \tag{5.3}$$

which is proved simply by noticing that the operators D / Dt and $\partial / \partial X^K$ commute.

As an example by using (5.3) we calculate the material derivatives:

$$\dot{C}_{KL} = \frac{D}{Dt} (\delta_{k\ell}\, x^k_{,K}\, x^\ell_{,L}) = 2d_{k\ell}\, x^k_{,K}\, x^\ell_{,L} \tag{5.4}$$

$$\dot{J} = \frac{D}{Dt} (\det x^k_{,K}) = \frac{\partial}{\partial x^k_{,K}} \det (x^\ell_{,L}) \frac{D}{Dt} (x^k_{,K})$$

$$= (JX^K_{,k})(v^k_{,\ell}\, x^\ell_{,K}) = J\, v^k_{,k} \tag{5.5}$$

where $d_{k\ell}$ represent the rate of deformation so that

$$d_{k\ell} \equiv \frac{1}{2} (v_{k,\ell} + v_{\ell,k}) \tag{5.6}$$

is called the *deformation rate tensor* which may be used to calculate

$$\frac{D}{Dt} (ds)^2 = 2d_{k\ell}\, dx^k\, dx^\ell$$

$$\frac{D}{Dt} (dv) = v^k_{,k}\, dv \tag{5.7}$$

From (5.7) it is clear that

Theorem (Killing). The necessary and sufficient condition for the motion to be locally rigid is $d_{k\ell} = 0$.

Useful for the measure of the rate of rotation is the *spin tensor* $w_{k\ell}$ defined by

$$w_{k\ell} \equiv \frac{1}{2} (v_{k,\ell} - v_{\ell,k}) \tag{5.8}$$

The corresponding axial vector defined by

$$w^k = \varepsilon^{k\ell m}\, w_{m\ell} \quad \text{or} \quad \underset{\sim}{w} = \nabla \times \underset{\sim}{v} \tag{5.9}$$

is called the *vorticity vector*.

Extensive accounts exist in the kinematics of continua and for these the reader is referred to Eringen [1962, 1967, 1975], Truesdell and Toupin [1962].

6. BALANCE LAWS

In continuum physics the following eight balance laws are postulated to be valid, irrespective of material constitution and geometry

(i) Conservation of Mass

(ii) Balance of Momentum

(iii) Balance of Moment of Momentum

(iv) Conservation of Energy

(v) Entropy Inequality (Second law of thermodynamics)

(vi) Conservation of Charge

(vii) Faraday's Law

(viii) Ampère's Law

For electromagnetically neutral media the last three laws may be disregarded. The first six laws can all be represented in the common form

$$\frac{D}{Dt}\int_{V-\sigma} \phi \, dv - \oint_{S-\sigma} \underset{\sim}{\tau} \cdot d\underset{\sim}{a} - \int_{V-\sigma} g \, dv = 0 \qquad (6.1)$$

except for the entropy inequality where (=) sign is replaced in (6.1) by ($\geq$). Here ϕ is a tensor field whose body source is g and the surface flux is $\underset{\sim}{\tau}$. Thus (6.1) expresses that *the time rate of change of the total field in the body* (with volume V excluding a surface of discontinuity σ that may be sweeping the body with a velocity $\underset{\sim}{\nu}$) *is equal* (in the case of (v) is greater than or equal) *to the field* $\underset{\sim}{\tau}$ *that comes into the body through its surface* $S-\sigma$ *and the source* g *that exists in* $V - \sigma$

For the electromagnetic continua (6.1) is supplemented by the laws

(vii) and (viii) which are expressible in the common form

$$\frac{D}{Dt}\int_{S-\sigma} \mathbf{q}\cdot d\mathbf{a} - \int_{C-\sigma} \mathbf{h}\cdot d\mathbf{s} - \int_{S-\sigma} \mathbf{r}\cdot d\mathbf{a} = 0 \tag{6.2}$$

where S is a *closed, two-sided material surface* passing through the *regular closed boundary curve* C · $S-\sigma$, and $C-\sigma$ represent, respectively, S and C excluding the points of interactions of S and C with the moving discon - tinuity surface σ. Thus (6.2) expresses *the balance of the flux of the total field* $\mathbf{q}$ *present on* $S-\sigma$ *with the source,* $\mathbf{r}$, *of* $\mathbf{q}$ *on* $S-\sigma$ *and the flux* $\mathbf{h}$ *of* $\mathbf{q}$ *which comes in through* $C-\sigma$. See Figs. 3 and 4.

We have the identifications

Faraday's law (6.3)

$$\mathbf{q} = \frac{1}{c}\mathbf{B}, \qquad \mathbf{h} = \mathbf{E} + \frac{1}{c}\mathbf{v}\times\mathbf{B}, \qquad \mathbf{r} = 0$$

Ampere's law

$$\mathbf{q} = -\frac{1}{c}\mathbf{D}, \qquad \mathbf{h} = \mathbf{H} - \frac{1}{c}\mathbf{v}\times\mathbf{D}, \qquad \mathbf{r} = \mathbf{J} - \rho_e \mathbf{v}$$

Since V in (6.1) is a material volume the operator D / Dt commutes with the integral sign. Using (5.7) and the Green-Gauss theorem (6.1) may be transformed to

$$\int_{V-\sigma} [\frac{\partial\phi}{\partial t} + \operatorname{div}(\phi\mathbf{v}) - \operatorname{div}\boldsymbol{\tau} - g]dv + \int_{\sigma} [\phi(\mathbf{v}-\boldsymbol{\nu})-\boldsymbol{\tau}]\cdot\mathbf{n}da = 0 \tag{6.4}$$

Here the integral on σ arises from the discontinuity surface which may be sweeping the body with the velocity $\boldsymbol{\nu}$ in the direction of its unit normal $\mathbf{n}$.

Localization of (6.4) is achieved by simply writing it in the equivalent form

$$\frac{\partial\phi}{\partial t} + \operatorname{div}(\phi\mathbf{v} - \boldsymbol{\tau}) - g = \hat{\phi} \qquad \text{in } V - \sigma \tag{6.5}$$

$$[\phi(\mathbf{v} - \boldsymbol{\nu}) - \boldsymbol{\tau}]\cdot\mathbf{n} = \hat{\Phi} \qquad \text{on } \sigma$$

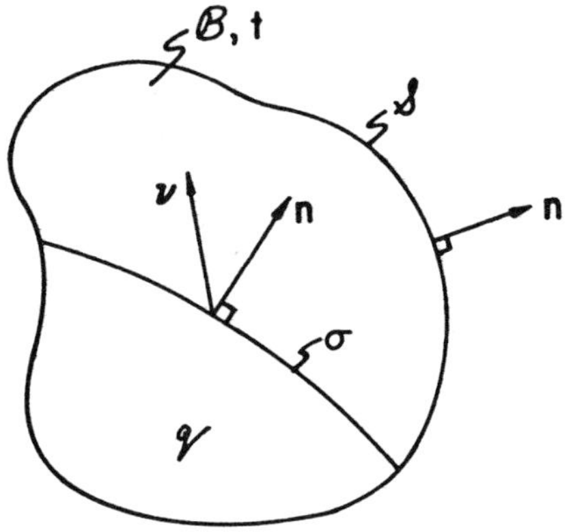

Fig. 3 Moving Discontinuity Surface σ

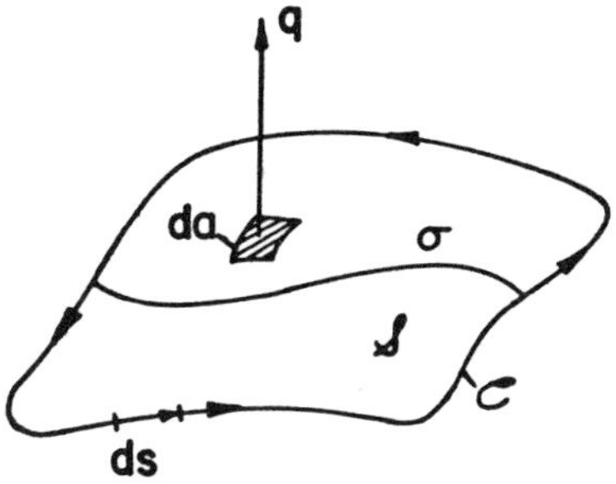

Fig. 4 Material Surface S with a Moving Discontinuity Line σ

subject to

$$\int_{V-\sigma} \hat{\phi}\, dv + \int_{\sigma} \hat{\Phi}\, da = 0 \tag{6.6}$$

These equations are exact and, as we shall see, they are valid for polar and nonlocal theories. Classical continuum mechanics makes its first departure here by postulating

Axiom of locality. The global balance loads (6.4) is posited to be valid for every part of the body. This means that in (6.4), V and σ can be replaced by any arbitrarily small volume $v \in V$ and $\sigma' \in \sigma$. It is then a consequence of a mathematical theorem that

$$\hat{\phi} = \hat{\Phi} = 0 \tag{6.7}$$

This postulate will be lifted when we come to discuss nonlocal field theories.

For E-M theories the approach is parallel to this but uses (6.2).

Employing (6.5) we obtain the local balance laws. To this end we identify ϕ, $\underset{\sim}{\tau}$ and g fields. They are listed below

		Fields	
Laws	ϕ	$\underset{\sim}{\tau}$	g
(i)	ρ	0	0
(ii)	$\rho\underset{\sim}{v}$	$\underset{\sim}{t}^k$	$\rho\underset{\sim}{f}$
(iii)	$\underset{\sim}{p} \times \rho\underset{\sim}{v}$	$\underset{\sim}{p} \times \underset{\sim}{t}^k$	$\underset{\sim}{p} \times \rho\underset{\sim}{f}$
(iv)	$\rho(\varepsilon + \frac{1}{2}\underset{\sim}{v} \cdot \underset{\sim}{v})$	$\underset{\sim}{t}^k \cdot \underset{\sim}{v} + q^k$	$\rho\, \underset{\sim}{f} \cdot \underset{\sim}{v} + \rho h$
(v)	$\rho\eta$	q^k / θ	$\rho h / \theta$

where

ρ	$\equiv$	mass density	,	$\underset{\sim}{v}$	$\equiv$	velocity vector	,
$\underset{\sim}{t}^k$	$\equiv$	stress vectors	,	$\underset{\sim}{f}$	$\equiv$	body force density	,
ε	$\equiv$	internal energy density	,	$\underset{\sim}{q}$	$\equiv$	the heat vector	,
η	$\equiv$	entropy density	,	θ	$\equiv$	absolute temperature	.

The stress vectors have the decomposition

$$\underset{\sim}{t}^k = t^{k\ell}\, i_\ell \tag{6.8}$$

where $t^{k\ell}$ is the *stress tensor*

Employing the above table in (6.5), it is straightforward to arrive at the local balance laws.

(i) *Mass*

$$\frac{\partial\rho}{\partial t} + (\rho\, v^k)_{,k} = 0 \qquad \text{in } V - \sigma$$

$$[\rho(\underset{\sim}{v} - \underset{\sim}{\nu})] \cdot \underset{\sim}{n} = 0 \qquad \text{on } \sigma \tag{6.9}$$

(ii) *Momentum*

$$t^{k\ell}{}_{,k} + \rho(f^\ell - \dot{v}^\ell) = 0 \qquad \text{in } V - \sigma$$

$$[\rho v^\ell(v^k - \nu^k) - t^{k\ell}]n_k = 0 \qquad \text{on } \sigma \tag{6.10}$$

(iii) *Moment of Momentum*

$$t^{k\ell} = t^{\ell k} \qquad \text{in } V - \sigma \tag{6.11}$$

(iv) *Energy*

$$\rho\,\dot{\varepsilon} - t^{k\ell}\, v_{\ell,k} - q^k{}_{,k} - \rho h = 0 \qquad \text{in } V - \sigma$$

$$[(\rho\varepsilon + \tfrac{1}{2}\rho\underset{\sim}{v}\cdot\underset{\sim}{v})(v^k - \nu^k) - t^{k\ell}\, v_\ell - q^k]n_k = 0 \qquad \text{on } \sigma \tag{6.12}$$

(v) *Entropy*

$$\rho \dot{\eta} - \left(\frac{q^h}{\theta}\right)_{,k} - \frac{\rho h}{\theta} \geq 0 \qquad \text{in } V - \sigma$$

$$[\rho\, \eta(v^k - \nu^k) - \frac{q^h}{\theta}] n_k \geq 0 \qquad \text{on } \sigma \tag{6.13}$$

The first set of these expressions are the *balance laws* in V-σ and the second set are the *jump conditions* at σ. Note that the jump conditions for the moment of momentum balance are satisfied identically. This is an accident of the classical field theories. In polar theories actually (6.11) is replaced by a set of differential equations and separate jump conditions exist (Section 10), see also Eringen and Şuhubi [1964], Eringen [1970], Eringen and Kafadar [1975].

7. CONSTITUTIVE EQUATIONS

Balance laws are valid for all types of materials. The material characterization requires separate equations that must incorporate the material properties. A general theory of constitutive equations exists applicable to large class of materials: elastic solids, viscos fluids, memory dependent materials, including thermal and E-M effects. Here we mention briefly the main ideas for thermomechanical continua. Afterward we develop the equations of nonlinear elastic solids (Section 8), Stokesian fluids (Section 9) and viscoelastic materials (Section 10).

(i) *Axiom of Causality: All thermomechanical phenomena is the result of past motions and temperatures of all points of the body.*

(ii) *Axiom of Objectivity: Constitutive response functions must be form-invariant under time dependent rigid body motions.*

(iii) *Axiom of Admissibility: Constitutive equations must not violate the balance laws and the entropy inequality.*

There are other axioms regarding the material symmetry regulations and for approximations however, we need not dwell into these here (cf. Eringen [1966, 1967].

According to (i) we write *functional equations* for $t^{k\ell}$, q^k, ε and η, e.g.

$$t^{k\ell}(\underset{\sim}{X},t) = \hat{t}^{k\ell}[\underset{\sim}{x}(\underset{\sim}{X}',t'),\theta(\underset{\sim}{X}',t');\underset{\sim}{X},t] \; ; \; \underset{\sim}{X}' \in B \; , \; t' \leq t$$

Here $\underset{\sim}{\hat{t}}$ at $(\underset{\sim}{X},t)$ *is a functional over the past motions and temperatures of all material points* X' *of the body and a function of* $\underset{\sim}{X}$ *and* t. Similar functional equations are written for q^k, ε and η. Alternatively, it is convenient to use the *Helmholtz free energy* $\psi \equiv \varepsilon - \theta\,\eta$ in place of ε, with a scalar constitutive equation

$$\psi(\underset{\sim}{X},t) = \hat{\psi}[\underset{\sim}{x}(\underset{\sim}{X}',t'),\; \theta(\underset{\sim}{X}',t') \; ; \; \underset{\sim}{X},t] \; ; \quad \underset{\sim}{X}' \in B \; , \; t' \leq t \tag{7.1}$$

According to the axiom of objectivity ψ must be form-invariant under the transformations

$$\begin{aligned} &\bar{x}_k(\underset{\sim}{X}',\bar{t}') = Q_{k\ell}(t')\, x_\ell(\underset{\sim}{X}',t') + b_k(t') \\ &\bar{t}' = t' - a \quad , \quad \underset{\sim}{Q}\,\underset{\sim}{Q}^T = \underset{\sim}{Q}^T\,\underset{\sim}{Q} = \underset{\sim}{I} \quad , \quad \det \underset{\sim}{Q} = \pm 1 \end{aligned} \tag{7.2}$$

which represent rigid body motions (for det Q = +1) and time shift. Here $\{\underset{\sim}{Q}(t)\}$ is the full orthogonal group so that (7.2) also includes reflection of axis (det $\underset{\sim}{Q}$ = -1). Thus $\hat{\psi}$ is subject to

$$\hat{\psi}[\underset{\sim}{\bar{x}}(\underset{\sim}{X}',\bar{t}'),\; \theta(\underset{\sim}{X}',\bar{t}');\; \underset{\sim}{X},t] = \hat{\psi}[\underset{\sim}{x}(\underset{\sim}{X}',t'),\theta(\underset{\sim}{X}',t'),\underset{\sim}{X},t] \tag{7.3}$$

Now we examine (7.3) for the following three special cases in succession.

(a) $\underset{\sim}{Q} = \underset{\sim}{I},\ \underset{\sim}{b} = \underset{\sim}{0},\quad a = t$ (time shift)

(b) $\underset{\sim}{Q} = \underset{\sim}{I},\ \underset{\sim}{b} = \underset{\sim}{x}(\underset{\sim}{X},t')$ (translations)

(c) $\underset{\sim}{Q} = \underset{\sim}{Q},\ \underset{\sim}{b} = \underset{\sim}{0},\quad a = 0$ (rotations) which are equivalent to full use of (7.2). The result of (a) and (b) is

$$\psi(\underset{\sim}{X},t) = \hat{\psi}[\underset{\sim}{x}(\underset{\sim}{X}',t-\tau') - \underset{\sim}{x}(\underset{\sim}{X},t-\tau'),\ \theta(\underset{\sim}{X}',t-\tau');\ \underset{\sim}{X}] \tag{7.4}$$

$$\tau' \equiv t - t' \geq 0 \quad , \quad 0 \leq \tau' \leq \infty$$

and (c) places further restrictions

$$\begin{aligned}&\hat{\psi}\{\underset{\sim}{Q}(t-\tau')[\underset{\sim}{x}(\underset{\sim}{X}',t-\tau') - \underset{\sim}{x}(\underset{\sim}{X},t-\tau')],\ \theta(\underset{\sim}{X}',t-\tau');\ \underset{\sim}{X}\}\\ &\quad = \hat{\psi}\{\underset{\sim}{x}(\underset{\sim}{X}',t-\tau') - \underset{\sim}{x}(\underset{\sim}{X},t-\tau'),\ \theta(\underset{\sim}{X}',t-\tau');\ \underset{\sim}{X}\}\end{aligned} \tag{7.5}$$

Equation (7.4) subject to (7.5) is the most general constitutive equation for the free energy, valid for all materials. Consider now the following special cases:

(i) *Gradient Dependent Materials*. For brevity let us write (7.4) in the form

$$\psi(\underset{\sim}{X},t) = \hat{\psi}[\underset{\sim}{G}(\underset{\sim}{X}');\underset{\sim}{X}] \quad , \quad 0 \leq \tau' \leq \infty \tag{7.6}$$

$$\underset{\sim}{G}(\underset{\sim}{X}') \equiv \{\underset{\sim}{x}(\underset{\sim}{X}',t-\tau') - \underset{\sim}{x}(\underset{\sim}{X},t-\tau'),\ \theta(\underset{\sim}{X}',t-\tau')\}$$

Suppose that $\hat{\psi}$ is a *uniformly continuous* functional of $\underset{\sim}{G}(\underset{\sim}{X}')$, i.e. if $\underset{\sim}{G}_1(\underset{\sim}{X}')$ and $\underset{\sim}{G}_2(\underset{\sim}{X}')$ are any two functions in the class of functions G then for any $\varepsilon > 0$, there exists a quantity $\delta(\varepsilon)$ such that

$$|\hat{\psi}[\underset{\sim}{G}_1(\underset{\sim}{X}')] - \hat{\psi}[\underset{\sim}{G}_2(\underset{\sim}{X}')]| < \varepsilon \tag{7.7}$$

whenever

$$||\ \underset{\sim}{G}_1(\underset{\sim}{X}') - \underset{\sim}{G}_2(\underset{\sim}{X}')|| < \delta \tag{7.8}$$

where $||\quad||$ denote a properly defined *norm* (in (7.8) the *distance* between the two functions $\underset{\sim}{G}_1$ and $\underset{\sim}{G}_2$). For example,

$$\| \underset{\sim}{F}(\underset{\sim}{X}') \| = \sup \, [\underset{\sim}{F}(\underset{\sim}{X}') \cdot \underset{\sim}{F}(\underset{\sim}{X}')]^{\frac{1}{2}} \tag{7.9}$$

or[1]

$$\| \underset{\sim}{F}(\underset{\sim}{X}') \| = \int_{V-\Sigma} H(|\underset{\sim}{X}'-\underset{\sim}{X}|)\underset{\sim}{F}(\underset{\sim}{X}') \cdot \underset{\sim}{F}(\underset{\sim}{X}') \, dV(\underset{\sim}{X}') \tag{7.10}$$

where $H(|\underset{\sim}{X}'|)$ (called *alleviator*) is a decreasing function of $|X'|$ with $H(0) = 1$. If $\underset{\sim}{G}(\underset{\sim}{X}')$ is expressible as a convergent Taylor series about $\underset{\sim}{X}' = \underset{\sim}{X}$ then according to a theorem of Weirstrass it may be approximated with any desired degree of accuracy by including a sufficiently large number of terms in the series

$$\underset{\sim}{G}^{(M)}(\underset{\sim}{X}') = \sum_{\mu=0}^{M} \frac{1}{\mu!} (X'^{K_1} - X^{K_1}) \ldots (X'^{K_\mu} - X^{K_\mu}) G_{,K_1 \ldots K_\mu}(\underset{\sim}{X}) \tag{7.11}$$

The functional $\hat{\psi}[\underset{\sim}{G}(\underset{\sim}{X}')]$ may now be uniformly approximated by a function of $\underset{\sim}{G}_{,K_1 \ldots K_\mu}$ (still being a functional with respect to τ'), with $\mu = 0,1,2,\ldots,M$.

(ia) *Simple Solids*. The case for which $M = 1$ is called *simple materials*. These materials are *local* in character. Thus for the simple memory-dependent materials the most general form of $\hat{\psi}$ is

$$\psi(\underset{\sim}{X},t) = \hat{\psi}[\underset{\sim}{x}_{,K}(t-\tau'), \ \theta(t-\tau'), \ \theta_{,K}(t-\tau'); \ \underset{\sim}{X}] \tag{7.12}$$

where and henceforth we avoid indicating the dependence of the argument functions on $\underset{\sim}{X}$, for brevity.

The axiom of objectivity places restrictions on the form (7.12). A theorem in the theory of invariants states that a scalar function $\hat{\psi}$ of vectors $\underset{\sim}{x}_{,K}$ will be objective if and only if it depends on the scalar products of $\underset{\sim}{x}_{,K}$, i.e.

[1]The choice of the norm depends on the nature of the intermolecular forces.

$$\psi(\underset{\sim}{X},t) = \hat{\psi}[C_{KL}(t-\tau'), \theta(t-\tau'), \theta_{,K}(t-\tau'); \underset{\sim}{X}] \tag{7.13}$$

where C_{KL} is the Green-deformation tensor. Equation (7.13) constitutes the basis of all phenominological theories concerning thermo-viscoelastic solids.

(ib) *Thermo-elastic Solids.* The elastic solids do not depend on the memory of past motions. Hence $\hat{\psi}$ for the elastic solids is a *function* of the form

$$\psi(\underset{\sim}{X},t) = \hat{\psi}(C_{KL}, \theta, \theta_{,K}, \underset{\sim}{X}) \tag{7.14}$$

(ii) *Simple Fluids.* To derive the general constitutive equations of simple fluids one must start with the relative motion $\underset{\sim}{x}_{(t)}(\tau)$ in place of $\underset{\sim}{x}(t)$. In exactly the same way this leads to

$$\psi(\underset{\sim}{x},t) = \hat{\psi}[\rho_{(t)}(t-\tau'), \underset{\sim}{c}_{(t)}(t-\tau'), \theta(t-\tau'), \theta_{,k}(t-\tau')] \tag{7.15}$$

where the argument functions depend on $\underset{\sim}{x}$ but $\hat{\psi}$ does not depend on $\underset{\sim}{x}$ explicitly, on account of objectivity. We draw attention to the inclusion of the relative density $\rho_{(t)}(\tau)$ in the argument. This is the result of the extra-property of the fluids that, for fluids the group of transformation $Q(t)$ are *unimodular* or *volume preserving* (physically, fluids possess pressure at their natural states).

Equation (7.15) is the basis of all thermo-viscous fluids such as polymeric fluids.

(iia) *Rate-Dependent Fluids.* If $\hat{\psi}$ is uniformly continuous in its argument functions, in the sense described in (i), and the argument functions accept Taylor series expansion about $\tau' = 0$, again we can replace the functional $\hat{\psi}$ by a function involving time rates, of $\underset{\sim}{c}_{(t)}$, θ, $\theta_{,k}$, e.g.

$$\underset{\sim}{c}^{(\mu)}_{(t)}(t) = \partial^{\mu}\, \underset{\sim}{c}_{(t)} \,/\, \partial\tau'^{\mu}\Big|_{\tau'=0} \quad , \quad \mu = 0, 1, \ldots, M$$
$$\partial^{0}\, \underset{\sim}{c}_{(t)} \,/\, \partial\tau'^{0}\Big|_{\tau'=0} = \underset{\sim}{c}_{(t)}(t) = 1 \tag{7.16}$$

(iib) *Stokesian Fluids.* Stokesian thermo-viscous fluids are obtained by retaining only the lowest order rates of the argument function in (7.15). Thus

$$\psi(\underset{\sim}{x},t) = \hat{\psi}(\rho^{-1}, d_{k\ell}, \theta, \theta_{,k}) \tag{7.17}$$

since $\dot{\underset{\sim}{c}}_{(t)}(t) = \underset{\sim}{d}$.

These examples are sufficient to show the origin of various types of continuum field theories.

We have yet to see the consequence of the axiom of thermodynamic admissibility. This is carried out, in some detail, for elastic solids in Section 8, for viscous fluids in Section 9 and for viscoelastic materials in Section 10. For more detailed accounts on the subject see Truesdell and Noll [1965], Eringen [1967], Dill [1974] and Rivlin [1974].

8. THEORY OF ELASTICITY

The free energy for the thermoelastic solids are given by (7.14). We consider here the case of non-heat conducting elastic solids. For thermoelastic solids, we refer the reader to Eringen [1967], Suhubi [1975]. In this case ψ is independent of $\theta_{,K}$ and we have

$$\psi(\underset{\sim}{X},t) = \hat{\psi}(C_{KL}, \theta, \underset{\sim}{X}) \tag{8.1}$$

Constitutive equations for $t^{k\ell}$, q^{k} and η also depend on the same argument as in (8.1). To see the consequence of entropy inequality we first eliminate h between $(6.12)_1$ and $(6.13)_1$ leading to

$$-\frac{\rho}{\theta}(\dot{\psi}+\dot{\theta}\eta)+\frac{1}{\theta}t^{k\ell}v_{\ell,k}+\frac{1}{\theta^2}q^k\theta_{,k}\geqq 0 \tag{8.2}$$

where we also used $\varepsilon = \psi + \theta\eta$. Calculating $\dot{\psi}$ from (8.1) and substituting into (8.2) we will have

$$-\frac{\rho}{\theta}\left(\frac{\partial\hat{\psi}}{\partial\theta}+\eta\right)\dot{\theta}+\frac{1}{\theta}\left(t^{k\ell}-2\rho\frac{\partial\hat{\psi}}{\partial C_{KL}}x^k_{,K}x^\ell_{,L}\right)d_{k\ell}+\frac{1}{\theta^2}q^k\theta_{,k}\geqq 0 \tag{8.3}$$

where we used (5.4). This inequality must not be violated for *all* independent variations of $\dot{\theta}$, $\underset{\sim}{d}$, and $\theta_{,k}$. But (8.3) is linear in these quantities since ψ, $\underset{\sim}{t}$, η, and $\underset{\sim}{q}$ do not depend on them. Therefore, the necessary and sufficient conditions that (8.3) will not be violated are

$$\eta=-\frac{\partial\hat{\psi}}{\partial\theta}\quad,\qquad t^{k\ell}=2\rho\frac{\partial\hat{\psi}}{\partial C_{KL}}x^k_{,K}x^\ell_{,L}\quad,\qquad q^k=0 \tag{8.4}$$

Hence there will be no heat conduction. Here we see the strength of the second law of thermodynamics in arriving at these elegant results. Given $\hat{\psi}(\underset{\sim}{C}, \theta, \underset{\sim}{X})$ then the constitutive equations are fully determined.

These results are exact and valid for finite deformations and temperatures and for inhomogeneous and anisotropic bodies. Various approximate theories (e.g. linear and quadratic theories in strains) can be constructed by merely writing power series expansion of $\hat{\psi}$ in C_{KL}. For these see Eringen [1962, 1967].

Material Symmetry. Materials possess various symmetry regulations. Such symmetries mathematically expressible by the invariance of $\hat{\psi}$ under a group of transformations of the material frame of reference, in the form

$$\begin{aligned}&\bar{\underset{\sim}{X}}=\underset{\sim}{S}\,\underset{\sim}{X}+\underset{\sim}{B}\\&\underset{\sim}{S}\,\underset{\sim}{S}^T=\underset{\sim}{S}^T\underset{\sim}{S}=\underset{\sim}{I}\quad,\qquad\det\underset{\sim}{S}=\pm 1\end{aligned} \tag{8.5}$$

where $\underset{\sim}{B}$ is a vector which is a *inhomogeneity* indicator while the group $\{S\}$ is a *symmetry group*. Under (8.5), (8.1) is subject to

$$\hat{\psi}(\underset{\sim}{C}, \theta, \underset{\sim}{X}) = \hat{\psi}(\underset{\sim}{S}\,\underset{\sim}{C}\,\underset{\sim}{S}^T, \theta, \underset{\sim}{S}\,\underset{\sim}{X} + \underset{\sim}{B}) \tag{8.6}$$

These are the restrictions arising from the material symmetry. When $\hat{\psi}$ is independent of $\underset{\sim}{B}$ (hence it does not depend on $\underset{\sim}{X}$ explicitly) the material is called *homogeneous*. When $\{S\}$ is the full group it is called *isotropic*. The entire 32 classes of crystalline materials are included in the 11 subgroups of $\{S\}$ and one reflection of X^K-axes.

Here we give only the specific results for the isotropic materials, for various classes of anisotropic materials, see Green and Adkins [1960], Suhubi [1975].

Isotropic Elastic Materials. For isotropic materials $\{S\}$ is the full group. Restriction (8.6) then implies that $\hat{\psi}$ must depend only on the invariants of $\underset{\sim}{C}$, i.e.

$$I_1 \equiv \text{tr}\,\underset{\sim}{C} = \text{tr}\,\overset{-1}{\underset{\sim}{c}}\quad, \quad I_2 \equiv \text{tr}\,\underset{\sim}{C}^2 = \text{tr}\,\overset{-2}{\underset{\sim}{c}}\quad, \quad I_3 \equiv \text{tr}\,\underset{\sim}{C}^3 = \text{tr}\,\overset{-3}{\underset{\sim}{c}} \tag{8.7}$$

Introducing the stress potential $\Sigma \equiv \rho_0\,\hat{\psi}$ and using (8.4) we obtain

$$t^{k\ell} = 2\,\frac{\rho}{\rho_0}\,\frac{\partial \Sigma}{\partial I_\alpha}\,\frac{\partial I_\alpha}{\partial C_{KL}}\,x^k_{,K}\,x^\ell_{,L}$$

If this differentiation is carried out and Cayley-Hamilton theorem is used we obtain

$$t^{k\ell} = a_o\,\delta^{k\ell} + a_1\,\overset{-1}{c}{}^{k\ell} + a_2\,\overset{-1}{c}{}^{k}{}_{m}\,\overset{-1}{c}{}^{m\ell} \tag{8.8}$$

where

$$\begin{aligned}
a_o &\equiv \frac{\rho}{\rho_0}\,(I_1^3 - 3I_1\,I_2 + 2I_3)\,\frac{\partial \Sigma}{\partial I_3} \\
a_1 &\equiv \frac{\rho}{\rho_0}\,[\,2\,\frac{\partial \Sigma}{\partial I_1} - 3(I_1^2 - I_2)\,\frac{\partial \Sigma}{\partial I_3}\,] \\
a_2 &\equiv \frac{\rho}{\rho_0}\,(4\,\frac{\partial \Sigma}{\partial I_2} + 6\,\frac{\partial \Sigma}{\partial I_3}\,)
\end{aligned} \tag{8.9}$$

From (8.8) one can derive various approximate theories (e.g. linear theory, quadratic theory, incompressible solids, etc.). Other forms of (8.9) involving $c_{k\ell}$, $e_{k\ell}$, etc. exist cf. Eringen [1962].

Finite elasticity have been used to solve various problems. It has found applications especially in rubber elasticity where the strains by nature are finite and therefore the linear theory is useless. For some of these solutions see Green and Zerna [1954], Green and Adkins [1960], Eringen [1962], and Truesdell and Noll [1965].

9. STOKESIAN FLUIDS

The constitutive equations for non-heat conducting Stokesian fluids, according to (7.17), are

$$\psi(\underset{\sim}{x},t) = \hat{\psi}(\rho^{-1}, d_{k\ell}, \theta) \tag{9.1}$$

and similar equations are written for $t^{k\ell}$, q^k, and η. The second law of thermodynamics (8.2) now reads

$$-\frac{\rho}{\theta}\left(\frac{\partial\hat{\psi}}{\partial\theta} + \eta\right)\dot{\theta} - \frac{\rho}{\theta}\frac{\partial\hat{\psi}}{\partial d_{k\ell}}\dot{d}_{k\ell} + \frac{1}{\theta}\,{}_Dt^{k\ell}\,d_{\ell k} + \frac{q^k}{\theta^2}\theta_{,k} \geq 0 \tag{9.2}$$

where ${}_D\underset{\sim}{t}$ is the *dissipative stress* defined by

$${}_Dt^{k\ell} \equiv t^{k\ell} + \pi\delta^{k\ell} \qquad ; \qquad \pi \equi - \frac{\partial\psi}{\partial\rho^{-1}} \tag{9.3}$$

The inequality (9.2) is not violated for all *independent* variations of $\dot{\theta}$, $\dot{d}_{k\ell}$, and $\theta_{,k}$, if and only if

$$\eta = -\frac{\partial\hat{\psi}}{\partial\theta} \quad , \quad \frac{\partial\hat{\psi}}{\partial d_{k\ell}} = 0 \quad , \quad q^k = 0 \tag{9.4}$$

$$\frac{1}{\theta}\,{}_Dt^{k\ell}\,d_{\ell k} \geq 0 \tag{9.5}$$

Thus $\hat{\psi}$ is independent of $\underset{\sim}{d}$ and no heat conduction takes place ($\underset{\sim}{q} = \underset{\sim}{0}$). Moreover, if ${}_D\underset{\sim}{t}$ is continuous in $\underset{\sim}{d}$ from (9.5), it follows that

$${}_D\underset{\sim}{t} = 0 \qquad \text{when } \underset{\sim}{d} = \underset{\sim}{0} \tag{9.6}$$

According to the axiom of objectivity ${}_D\underset{\sim}{t}$ must obey

$$\underset{\sim}{Q}\ {}_{D}\underset{\sim}{t}(\rho^{-1}, \underset{\sim}{d}, \theta)\underset{\sim}{Q}^{T} = {}_{D}\underset{\sim}{t}(\rho^{-1}, \underset{\sim}{Q}\ \underset{\sim}{d}\ \underset{\sim}{Q}^{T}, \theta) \tag{9.7}$$

for all Q subject to $(7.2)_3$. This means that ${}_{D}\underset{\sim}{t}$ is an isotropic tensor. Hence we have

Theorem. Stokesian fluids are isotropic.

If ${}_{D}\underset{\sim}{t}$ is uniformly continuous then according to a theorem of Weierstrass it can be approximated by a polynomial in $\underset{\sim}{d}$ to any desired degree. Thus writing

$${}_{D}\underset{\sim}{t} = \sum_{m=0}^{M} \beta_m(\rho^{-1}, \theta)\ \underset{\sim}{d}^{m}$$

and using Cayley-Hamilton theorem we can express all terms containing d^m, $m \geq 3$ in terms of $\underset{\sim}{I}$, $\underset{\sim}{d}$, and $\underset{\sim}{d}^2$. Thus

$${}_{D}\underset{\sim}{t} = b_o\underset{\sim}{I} + b_1\underset{\sim}{d} + b_2\underset{\sim}{d}^2 \tag{9.8}$$

where b_α are polynomials in the invariants of $\underset{\sim}{d}$ and depend on ρ^{-1} and θ, i.e.

$$b_\alpha = \hat{b}_\alpha(\rho^{-1}, \theta, \mathrm{tr}\ \underset{\sim}{d}, \mathrm{tr}\ \underset{\sim}{d}^2, \mathrm{tr}\ \underset{\sim}{d}^3) \quad , \quad \alpha = 0, 1, 2 \tag{9.9}$$

Thus again we have the elegant result

$$t^{k\ell} = (-\pi + b_o)\delta^{k\ell} + b_1\ d^{k\ell} + b_2\ d^{k}_{\ m}\ d^{m\ell} \tag{9.10}$$

where the thermodynamic pressure $\pi(\rho^{-1}, \theta)$ is given by $(9.3)_2$. Of course ${}_{D}\underset{\sim}{t}$ given by (9.8) must further be restricted by the entropy inequality (9.5). When (9.10) is linearized with respect to $\underset{\sim}{d}$ we obtain the celebrated Navier-Stokes theory.

10. VISCOELASTICITY

Constitutive equations of viscoelastic materials are of the form (7.13) that must be restricted by the second law of thermodynamics. Here we consider only the isothermal case for simplicity, thus

$$\psi(\underset{\sim}{X},t) = \hat{\psi}[\underset{\sim}{C}(t-\tau'); \theta, \underset{\sim}{X}] \quad , \quad 0 \leq \tau' < \infty \tag{10.1}$$

To include the viscous fluids and Kelvin-Voight solids, one needs to enlarge the domain of τ' to $-\varepsilon \leq \tau' < \infty$ where ε is a small positive quantity that may be made to approach zero. Alternatively, one may include $\dot{\underset{\sim}{C}}$ into the argument of $\hat{\psi}$ and an additional variable.

We may write (10.1) also in the form

$$\psi(\underset{\sim}{X},t) = \hat{\psi}[\underset{\sim}{C}^X(t-\tau'); \underset{\sim}{C}, \theta, \underset{\sim}{X}] \; ; \quad \underset{\sim}{C}^X(t-\tau') \equiv \underset{\sim}{C}(t-\tau') - \underset{\sim}{C}(t) \tag{10.2}$$

and introduce a metric to the space of functions $\underset{\sim}{C}^X$ to provide appropriate differentiability requirement. Hilbert space H with an inner produce defined by

$$(\underset{\sim}{C}_1^X , \underset{\sim}{C}_2^X)_I \equiv \int_{-\infty}^{t} I(t-\tau') \, C_{1KL}^X(\tau') \, C_{2KL}^X(\tau') \, d\tau' \tag{10.3}$$

provides a proper norm $(\underset{\sim}{C}^X , \underset{\sim}{C}^X)_I^{1/2}$ to this end. Here positive decreasing function $I(\tau')$, with $I(0) = 1$, is the alleviator introduced to diminish the effect of strains at distant pasts in the values of the response functionals at $\tau' = t$. Viscoelasticity based on this norm is said to obey the *fading memory* hypothesis (Coleman & Noll [1961]).

We assume further that $\hat{\psi}$ is also continuously differentiable with respect to $\underset{\sim}{C}$ and θ. With this we can calculate

$$\dot{\psi} = \frac{\partial \hat{\psi}}{\partial \underset{\sim}{C}} \cdot \dot{\underset{\sim}{C}} + \frac{\partial \hat{\psi}}{\partial \theta} \dot{\theta} + \int_{-\infty}^{t} \frac{\delta \hat{\psi}}{\delta \underset{\sim}{C}^X} \cdot \dot{\underset{\sim}{C}}^X \, d\tau' \tag{10.4}$$

where $\delta\hat{\psi}/\delta\underset{\sim}{C}^X$ represents the Fréchet derivative of $\hat{\psi}$ with respect to $\underset{\sim}{C}^X$ with the norm defined above.

From the Clausius-Duhem inequality (8.2) it now follows (in the same way as in Section 8) that

$$\eta = -\frac{\partial \hat{\psi}}{\partial \theta}, \quad \underset{\sim}{q} = \underset{\sim}{0} ,$$
$$t^{k\ell} = \left(\rho_o \frac{\partial \hat{\psi}}{\partial C_{KL}} - \rho_o \int_{-\infty}^{t} \frac{\delta \hat{\psi}}{\delta C_{KL}^X} \, d\tau'\right) x^k_{,K} \, x^\ell_{,L} \tag{10.5}$$

and

$$\sigma \equiv - \frac{\rho_o}{\theta} \int_{-\infty}^{t} \frac{\delta \hat{\psi}}{\delta C^X_{\sim KL}} \dot{C}_{KL} (t - \tau') \, d\tau' \geq 0 \tag{10.6}$$

Here σ represents the *dissipation function* which must be non-negative for all independent processes.

These results possess many interesting thermodynamic consequences cf. Truesdell and Noll [1965], Eringen [1975]. By using various representations and/or approximations of functional $\hat{\psi}$ from (10.5), one can obtain various types of viscoelastic materials. For example, when $\hat{\psi}$ is a quadratic functional in $\underset{\sim}{C}^X$ (10.5) leads to Boltzman-Volterra type viscoelasticity with stress constitutive equations of the form

$$t^{k\ell} = \sigma^{k\ell mn} e_{mn} + \int_{-\infty}^{t} \gamma^{k\ell mn} (t - \tau') \frac{\partial e_{mn}(\tau')}{\partial \tau'} \, d\tau' \tag{10.7}$$

(cf. Eringen [1967, Section 9.6]).

Nonlinear viscoelasticity outlined above and corresponding memory dependent fluids have found extensive applications in the field of rheology. They are being used extensively for the problems concerning polymeric substances for some applications, see Lodge [1964], Coleman et al [1966], Eringen [1962, 1965], Lockett [1972], Dill [1975].

11. POLAR CONTINUA

Elements of polar continua are complex material points with two sets of degrees of freedom: (a) translation and (b) microdeformation. The motion of a material point in a *micromorphic continuum* is prescribed by

$$\underset{\sim}{x} = \hat{\underset{\sim}{x}}(\underset{\sim}{X}, t) \quad , \quad \underset{\sim}{\chi}_K = \hat{\underset{\sim}{\chi}}_K(\underset{\sim}{X}, t) \tag{11.1}$$

of which the first is the *macromotion* (or simply motion) and the second is the *micromotion*. The unique inverses of these are posited to exist:

$$\underset{\sim}{\chi} = \hat{\underset{\sim}{\chi}}(\underset{\sim}{x},t) \quad , \quad \underset{\sim k}{\chi} = \hat{\chi}_k(\underset{\sim}{x},t) \tag{11.2}$$

such that

$$\chi^k{}_K \; \chi^K{}_\ell = \delta^k{}_\ell \quad , \quad \chi^k{}_L \; \chi^K{}_k = \delta^K{}_L \tag{11.3}$$

When $\underset{\sim}{\chi}_K$ is an orthogonal matrix, i.e.

$$\underset{\sim}{\chi} = \underset{\sim}{\chi}^T \tag{11.4}$$

the medium is called *micropolar*. In this case the material points can undergo only translations and rigid rotations so that the elements of the body are like *small rigid particles*. Without the constraint (11.4) the material points are deformable so that $\underset{\sim}{\chi}$ describes both intrinsic rigid rotations and micro-shears and-stretches. In this case the body is called *micromorphic*. Here we present a brief account on micropolar elastic solids. For micropolar fluids and micromorphic continuum theories the reader is referred to Eringen and Şuhubi [1964], Eringen [1964, 1966, 1970, 1972] and Eringen and Kafadar [1971].

For micropolar continuum, we have the constraints (11.4) so that (11.3) reads

$$\chi^k{}_K \; \chi_\ell{}^K = \delta^k{}_\ell \quad , \quad \chi^k{}_L \; \chi_k{}^K = \delta^K{}_L \tag{11.5}$$

Material derivative of $\underset{\sim}{\chi}$ may be written as

$$\dot{\chi}_{kK} = \nu_{k\ell} \; \chi^\ell{}_K \tag{11.6}$$

where $\nu_{k\ell}(\underset{\sim}{x},t)$ is called the *gyration tensor* which can be solved from (11.6) by multiplying it by $\chi_m{}^K$:

$$\nu_{k\ell} = \dot{\chi}_{kK} \, \chi_\ell{}^K = - \nu_{\ell k} \tag{11.7}$$

which is skew symmetric on account of (11.5). Hence an axial vector ν^k may be introduced by

$$\nu^k = -\frac{1}{2}\varepsilon^{k\ell m}\nu_{\ell m} \quad , \quad \nu_{k\ell} = -\varepsilon_{k\ell m}\nu^m \tag{11.8}$$

Since now the elements of the body are "dipolar" in nature (or like small rigid bodies), they can carry intrinsic body and surface couples. Therefore, it is necessary to reformulate the balance laws. The laws of conservation of mass and balance of momentum remain unchanged but we must revise balance of moment of momentum and energy. Accordingly, we have

Law (iii):

$$\phi = \underset{\sim}{p} \times \rho \underset{\sim}{v} + \rho \underset{\sim}{\sigma} \quad , \quad \underset{\sim}{\tau} = \underset{\sim}{p} \times \underset{\sim}{t}^k + \underset{\sim}{m}^k \quad , \quad g = \underset{\sim}{p} \times \rho \underset{\sim}{f} + \rho \underset{\sim}{\ell}$$

Law (iv): (11.9)

$$\phi = \rho(\varepsilon + \frac{1}{2}\underset{\sim}{v}\cdot\underset{\sim}{v}) + \frac{1}{2}\rho\,\underset{\sim}{\sigma}\cdot\underset{\sim}{\nu} \quad , \quad \underset{\sim}{\tau} = \underset{\sim}{t}^k\cdot\underset{\sim}{v} + \underset{\sim}{m}^k\cdot\underset{\sim}{\nu} \quad , \quad g = \rho\underset{\sim}{f}\cdot\underset{\sim}{v} + \rho\underset{\sim}{\ell}\cdot\underset{\sim}{\nu}$$

where

$$\sigma_k = j_{k\ell}\,\nu^\ell \quad , \quad \underset{\sim}{m}^k = m^{k\ell}\,\underset{\sim}{i}_\ell \quad , \quad \underset{\sim}{\ell} = \ell^k\,\underset{\sim}{i}_k \tag{11.10}$$

Here $j_{k\ell}$ is the *microinertia density* per unit volume, $m^{k\ell}$ is the *couple stress* and ℓ^k is the *body couple density*. Thus in a micropolar body the material particles are endowed with *intrinsic angular momentum* $\underset{\sim}{\sigma}$, body couple $\underset{\sim}{\ell}$ and on the surface of the body there exists couple stress $m^{k\ell}$ in addition to the stress $t^{k\ell}$. See Fig. 5.

Upon carrying (11.9) and (11.10) into (6.5) with $\hat{\phi} = \hat{\Phi} = 0$, we obtain the new laws:

$$m^{k\ell}{}_{,k} + \varepsilon^{\ell mn}\,t_{mn} + \rho(\ell^\ell - \dot{\sigma}^\ell) = 0 \qquad \text{in } V - \sigma$$

$$[m^{k\ell} - \rho\sigma^\ell(v^k - u^k)]\,n_k = 0 \qquad \text{on } \sigma \tag{11.11}$$

$$-\rho\dot{\varepsilon} + t^{k\ell}(v_{\ell,k} + \nu_{k\ell}) + m^{k\ell}\,\nu_{\ell,k} + q^k{}_{,k} - \rho h = 0 \qquad \text{in } V - \sigma$$

$$[\underset{\sim}{t}^{k\ell} v_\ell + m^{k\ell}\nu_\ell - (\rho\varepsilon + \frac{1}{2}\rho\,\underset{\sim}{v}\cdot\underset{\sim}{v} + \frac{1}{2}\rho\,\underset{\sim}{\sigma}\cdot\underset{\sim}{\nu})(v^k - u^k)]\underset{\sim}{n}_k = 0 \qquad \text{on } \sigma \tag{11.12}$$

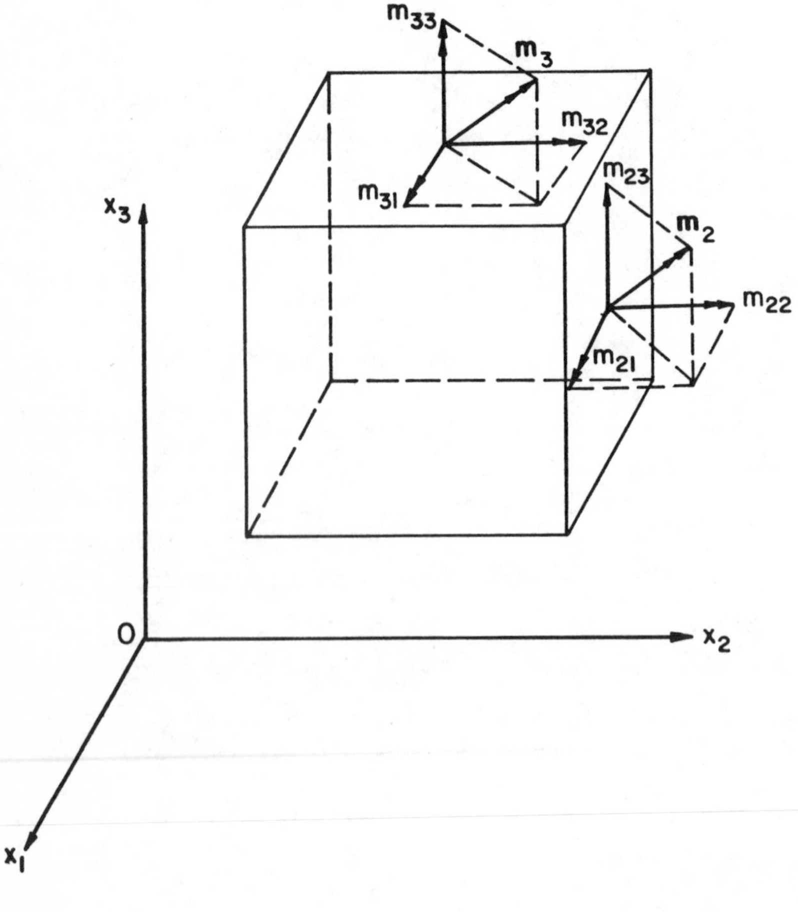

Fig. 5 Couple Stress

These two sets of equations replace respectively (6.11) and (6.12) of the classical theories.

Another set of balance laws is necessary for the microinertia tensor $j_{k\ell}$ and this was discovered by Eringen [1964]:

$$\frac{Dj^{k\ell}}{Dt} - j^{kr}\,\nu_r{}^{\ell} - j^{\ell r}\,\nu_r{}^{k} = 0 \tag{11.13}$$

It is now clear that in discussing the nature of micropolar bodies, we need to reexamine all basic concepts from the beginning and reconstruct the constitutive equations so that they will include the effect of the intrinsic rotatory deformations. Extensive accounts on the topic exist cf. Eringen and Şuhubi [1966, 1967, 1968, 1970], Eringen and Kafadar [1971].

Micropolar continuum theories are now well established. Several review articles exist, cf. Eringen [1968a], Stajanovic [1969], Ariman et al. [1973, 1974]. Since micropolar continuum theories constitute the basis of field theories for liquid crystals, we give here a very brief account on the micropolar elastic solids, only for a taste. We note, however, that without the polar theories, the entire field of liquid crystals and much of blood rheology and suspension mechanics remain without any organization and firm foundation.

12. MICROPOLAR ELASTICITY

The development of the general constitutive theory for micropolar media requires the inclusion of the past micromotions $\underset{\sim}{\chi}_K(\underset{\sim}{X}',t')$, of all points $\underset{\sim}{X}'$ of the body into the argument of the constitutive functionals. Thus ψ, $\underset{\sim}{t}$, $\underset{\sim}{m}$, $\underset{\sim}{q}$, and η are now functionals of $\underset{\sim}{x}(\underset{\sim}{X}',t')$, $\underset{\sim}{\chi}_K(\underset{\sim}{X}',t')$, $\theta(\underset{\sim}{X}',t')$, and a function of $\underset{\sim}{X}$ and t, for all $\underset{\sim}{X}'\varepsilon B$ and for all past times $t' \leq t$. A procedure similar to that followed in section 8.2 can be used to arrive at

$$\psi(\underset{\sim}{X},t) = \hat{\psi}[\underset{\sim}{x}_{,K}(t-\tau'),\ \underset{\sim}{\chi}_K(t-\tau'),\ \underset{\sim}{\chi}_{K,L}(t-\tau'),\ \theta(t-\tau'),\ \theta_{,K}(t-\tau'),\ \underset{\sim}{X}] \tag{12.1}$$

for the *simple micropolar solids* which are memory dependent. The axiom of objectivity then tells us that ψ can only depend on the scalar products of $\underset{\sim}{x}_{,K}$, $\underset{\sim}{\chi}_K$, and $\underset{\sim}{\chi}_{K,L}$ two at a time. In view of the conditions of orthogonality (11.5) we then find that all of these scalar products can be expressed in terms of the following two

$$\mathscr{C}_{KL} \equiv x^k_{,K}\,\chi_{kL} \quad , \quad \Gamma_{KL} \equiv \frac{1}{2}\,\varepsilon_{KMN}\,\chi^{kM}_{\ \ ,L}\,\chi_k^{\ N} \tag{12.2}$$

Of these $\mathscr{C}_{KL}$ is called *Cosserat deformation tensor* and Γ_{KL} the *Wryness tensor*. Thus (12.1) is objective if it is a functional of the form

$$\psi(\underset{\sim}{X},t) = \hat{\psi}[\mathscr{C}_{KL}(t-\tau'),\ \Gamma_{KL}(t-\tau'),\ \theta(t-\tau'),\ \theta_{,K}(t-\tau')\ ;\ \underset{\sim}{X}] \tag{12.3}$$

Subject to the axiom of admissibility, this is the form of the most general constitutive equations for the memory dependent micropolar materials.

For the micropolar elastic solids, there will be no memory dependence and for isothermal solids, no $\theta_{,K}$-dependence so that

$$\psi(\underset{\sim}{X},t) = \hat{\psi}(\mathscr{C}_{KL},\ \Gamma_{KL},\ \theta;\ \underset{\sim}{X}) \tag{12.4}$$

together with equations for $\underset{\sim}{t}$, $\underset{\sim}{m}$, $\underset{\sim}{q}$, and η containing the same arguments constitute the constitutive equations of the micropolar elastic solids. The Clausius-Duhem inequality that is obtained by eliminating h between (11.12) and (6.13), can then be used to arrive at

$$t^{k\ell} = \rho\,\frac{\partial\hat{\psi}}{\partial\mathscr{C}_{KL}}\,x^k_{,K}\,\chi^{\ell}_{\ L},\quad m^{k\ell} = \rho\,\frac{\partial\hat{\psi}}{\partial\Gamma_{LK}}\,x^k_{,K}\,\chi^{\ell}_{\ L},\quad \eta = -\frac{\partial\hat{\psi}}{\partial\theta}\ ,\quad q^k = 0 \tag{12.5}$$

From these exact equations, we can obtain all sorts of special cases. The material symmetry can be incorporated and approximate equations (linear, quadratic in strains, etc.) can be obtained.

Linear Theory. I reproduce here the linear theory for isotropic micropolar bodies since these may not be familiar to some readers

$$\begin{aligned} t_{k\ell} &= \lambda\, \varepsilon^{r}{}_{r}\, \delta_{k\ell} + (\mu + \kappa)\, \varepsilon_{k\ell} + \mu\, \varepsilon_{\ell k} \\ m_{k\ell} &= \alpha\, \gamma^{r}{}_{r}\, \delta_{k\ell} + \beta\, \gamma_{k\ell} + \gamma\, \gamma_{\ell k} \end{aligned} \tag{12.6}$$

where λ, μ, κ, α, β and γ are the material moduli, $\varepsilon_{k\ell}$ is the strain tensor and $\gamma_{k\ell}$ is the wryness tensor. For the linear theory we have

$$\begin{aligned} &\varepsilon_{k\ell} \simeq u_{\ell,k} - \varepsilon_{k\ell m}\, \phi^{m} \quad , \quad \gamma_{k\ell} \simeq \phi_{k,\ell} \\ &\nu^{k} \simeq \dot{\phi}^{k} \quad , \quad \nu_{k\ell} \simeq - \varepsilon_{k\ell m}\, \dot{\phi}^{m} \end{aligned} \tag{12.7}$$

where $\underset{\sim}{u}$ is displacement vector and $\underset{\sim}{\phi}$ is the rotation vector. Upon substituting (12.6) into the balance of momentum and momentum, we obtain the field equations.

$$\begin{aligned} &(\lambda + 2\mu + \kappa)\, \underset{\sim}{\nabla}\, \underset{\sim}{\nabla} \cdot \underset{\sim}{u} - (\mu{+}\kappa)\, \underset{\sim}{\nabla} \times \underset{\sim}{\nabla} \times \underset{\sim}{u} + \kappa\, \underset{\sim}{\nabla} \times \underset{\sim}{\phi} + \rho(\underset{\sim}{f} - \ddot{\underset{\sim}{u}}) = \underset{\sim}{0} \\ &(\alpha + \beta + \gamma)\, \underset{\sim}{\nabla}\, \underset{\sim}{\nabla} \cdot \underset{\sim}{\phi} - \gamma\, \underset{\sim}{\nabla} \times \underset{\sim}{\nabla} \times \underset{\sim}{\phi} + \kappa\, \underset{\sim}{\nabla} \times \underset{\sim}{u} - 2\kappa\, \underset{\sim}{\phi} + \rho(\underset{\sim}{\ell} - \underset{\approx}{j} \cdot \ddot{\underset{\sim}{\phi}}) = \underset{\sim}{0} \end{aligned} \tag{12.8}$$

The material will be thermodynamically stable if the free energy is non-negative for *all* motions. It can be shown that this will be the case if the material moduli is restricted by

$$\begin{aligned} &0 \le 3\lambda + 2\mu + \kappa \quad , \quad 0 \le 2\mu + \kappa \quad , \quad 0 \le \kappa \\ &0 \le 3\alpha + \beta + \gamma \quad , \quad 0 \le \gamma + \beta \quad , \quad 0 \le \gamma - \beta \end{aligned} \tag{12.9}$$

For these and other results, see Eringen [1970a]. A comparison of $(12.8)_1$ with Naviers equations (2.1) will show that in (12.8), there is a coupling term $\underset{\sim}{\nabla} \times \underset{\sim}{\phi}$ and of course $(12.8)_2$ is totally missing.

The use of the set (12.8) is essential if we are dealing with bodies whose elements are polar. Nature, of course, contains large numbers of such bodies. Granular solids, polymers, liquid crystals, suspensions, composites are but a few examples for such bodies. Several investigations exist indicating that the micropolar theory is applicable also to bodies consisting of certain types of molecular structure. In the following section, we discuss briefly two applications of the theory.

13. APPLICATIONS OF MICROPOLAR FIELD THEORIES

(i) *Plane Waves.* The plane wave solution of (12.8) leads to the following dispersion relations, Parfitt and Eringen [1969].

(a) *Longitudinal Waves*

$$\omega_{LA} = \left(\frac{\lambda+2\mu+\kappa}{\rho}\right)^{1/2} q \qquad (13.1)$$

$$\omega_{LO} = \omega_o\left(2 + \frac{\alpha+\beta+\lambda}{\kappa} q^2\right)^{1/2} \quad ; \quad \omega_o \equiv (\kappa/\rho j)^{1/2}$$

(b) *Transverse Waves*

$$\left.\begin{matrix}\omega_{TA} \\ \omega_{TO}\end{matrix}\right\} = \omega_o^{\,2} + \alpha_o^{\,2} q^2 \mp [(\omega_o^{\,2} + \alpha_o^{\,2} q^2)^2 - (\beta_o^{\,2} + \gamma_o^{\,2} q^2)q^2]^{1/2} \qquad (13.2)$$

$$\alpha_o \equiv [\gamma + (\mu+\kappa)\, j]/2\kappa \quad , \quad \beta_o \equiv \kappa(2\mu+\kappa)/\rho j$$

$$\gamma_o \equiv \gamma(\mu+\kappa)/\rho^2 j$$

where j is the microinertia for "microisotropic solids" ($j_{k\ell} = j\, \delta_{k\ell}$) $\omega/2\pi$ is the frequency and q is the wave number. The dispersion curves for

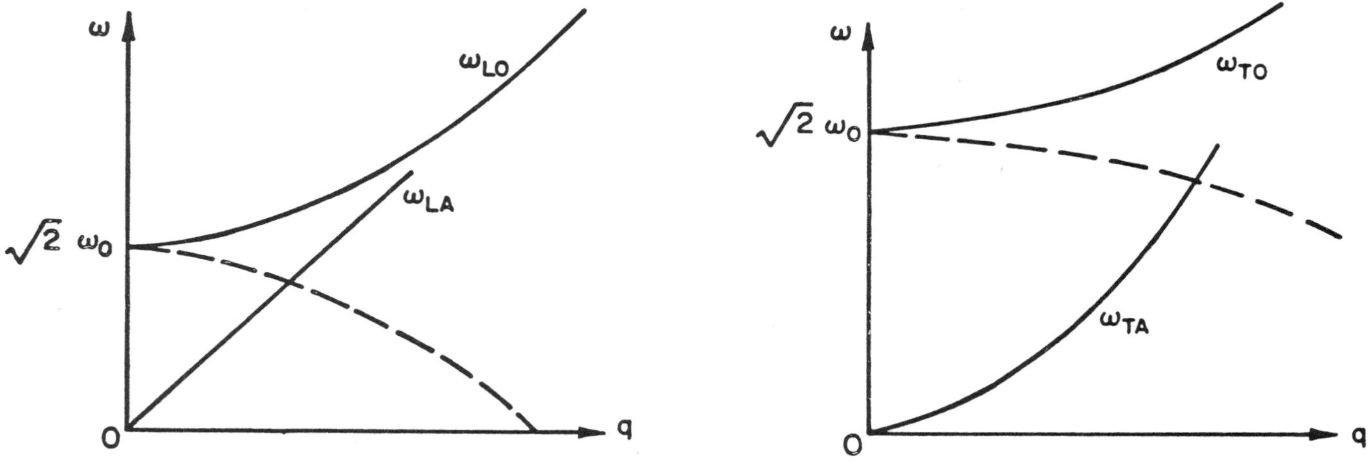

A. B.

Fig. 6 Dispersion of Plane Waves (Micropolar Theory)

A. Longitudinal Waves B. Transverse Waves

(13.1) are shown in Figs. 6A and 6B. We note the following properties:

α) Longitudinal acoustical modes (with frequency ω_{LA}) are non-dispersive just as in classical elasticity.

β) Longitudinal optical modes (with frequency ω_{LO}) are dispersive.

γ) Transverse Waves (which are totally missing in classical elasticity) are also dispersive. Depending on the material moduli ω_{LO} and ω_{TO} can increase or decrease with the wave number. At $q = 0$ (infinite wave length) we have a *cut-off* frequency of the optical modes

$$\omega_c = \sqrt{2}\,\omega_o = (2\kappa/\rho j)^{1/2} \tag{13.3}$$

This result can be used to calculate the new micropolar module κ. Here I give an example:

Potassium-Nitrate (KNO_3) has a typical molecule reminicent of large classes of materials. Experiments with Raman spectroscopy indicate that Balkanski and Teng [1969]:

$$\omega_c = 1.22 \times 10^{13} \text{ sec}^{-1}$$

Askar [1972] using this result and the molecular dimension of KNO_3 (Fig. 7) found that

$$\kappa/(\lambda+2\mu) \approx 1/10$$

The use of the micropolar model in justified because the resonance frequencies for rotational modes are 100 times higher than those for the stretch modes of the molecules. Note that the mere existence of these modes, as observed in experiments, disqualifies the classical elasticity theory as a model. However, micropolar theory and micromorphic theories appear to be excellent candidates for bodies where rotatory and stretch modes are important. Nevertheless, these theories too breakdown when we are dealing with wave

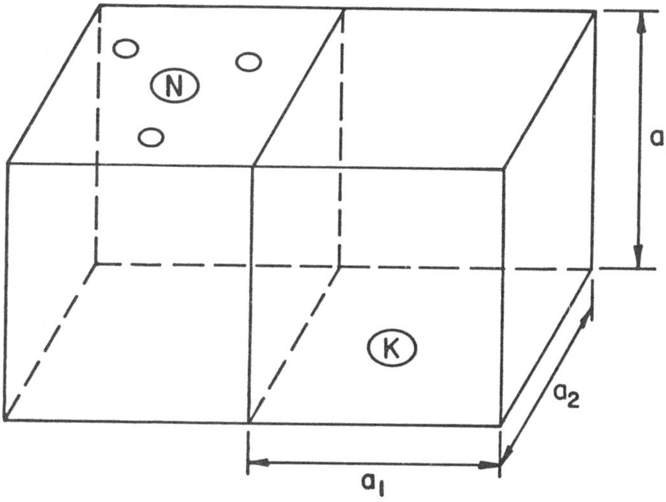

Fig. 7 KNO_3 MOLECULE

$a_1 \simeq 2.72$ Å , $a_2 \simeq 4.08$ Å , $a_3 \simeq 3.23$ Å

lengths that are smaller than the molecular dimensions (e.g. atomic dimensions) for which, as we shall see, it is necessary to resort to non-local field theories.

(ii) *Liquid Crystals*

For a large class of substances, known as liquid crystals, the classical continuum theory completely fails. These substances have cigar-shaped molecules (approximately of the order of 10^{-5} cm) arranged in parallel to each other and in layers. Schematic arrangements of three major classes of liquid crystals (nematic, smectic and cholesteric) are shown in Fig. 8. In nematic and smectic substances, the cigars are parallel to each other differing in their layer structure while cholesteric crystals form a helical structure in the direction perpendicular to layers. These substances exhibit many interesting physical phenomena not present in isotropic liquids and solids. For example, the viscosity of p-Azoxyanisole) a nematic substance, at 122°C is about 2.4, 9.2 and 3.4 centipoises when the molecules are respectively parallel to the flow direction, perpendicular to the velocity gradient and perpendicular to both. Cholosteric liquid crystals exhibit many interesting effects, e.g. birefringence, circular dichroism, optical activity and color displays. Because of the delicately balanced molecular structure, marked changes can be produced in response to subtle variations of temperature, mechanical stresses, E-M effects, etc. Already many industrial applications exist (e.g. digital thermometers, watches, display devices, memory devices). It is not possible here to discuss many intriguing physical phenomena nested in these materials. For some of these we refer the reader to Gray [1971] and de Gennes [1975].

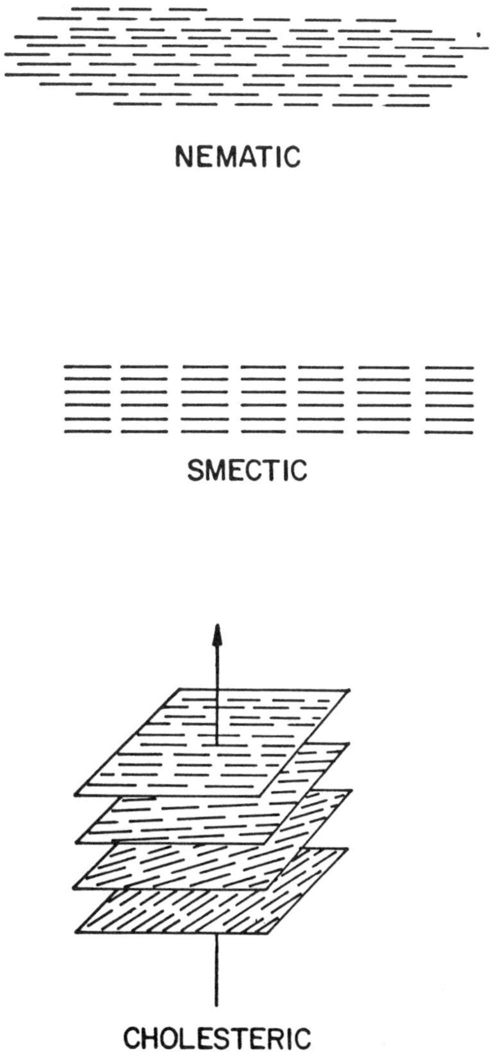

Fig. 8 Liquid Crystal Structures

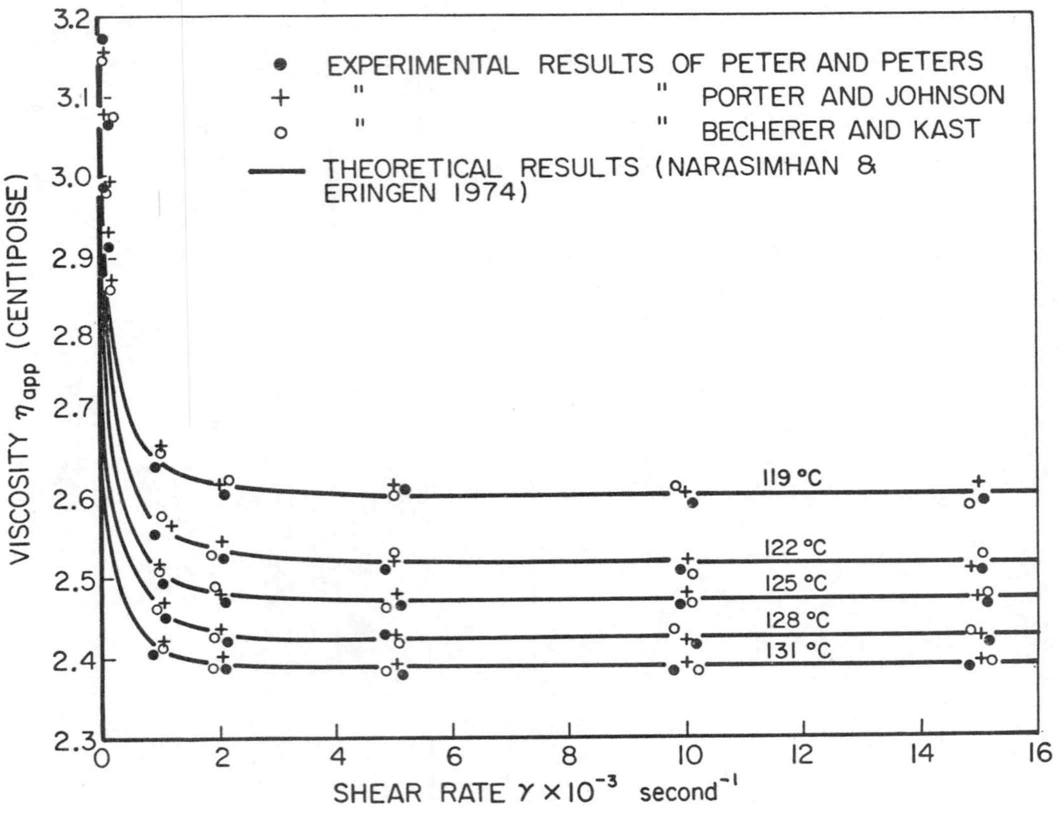

Fig. 9 Behavior of Apparent Viscosity with Shear-Rate

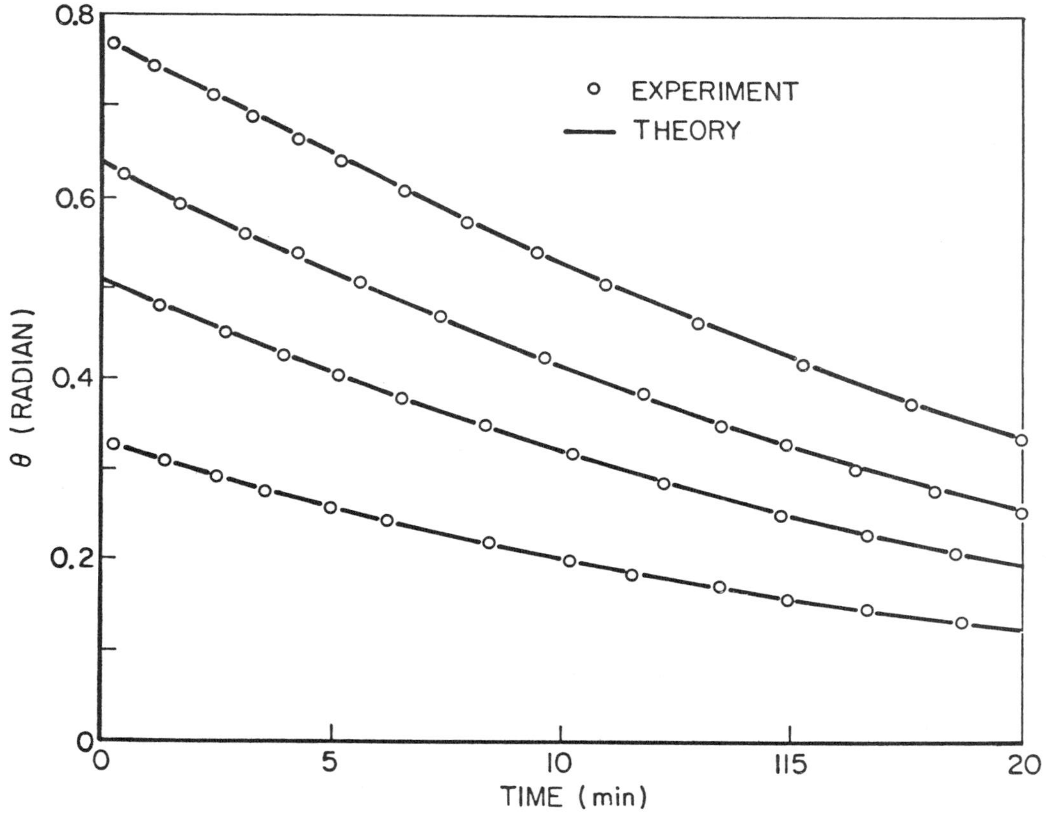

Fig. 10 Reorientation of a 16.8% (w/w) Solution of Racemic PBG in Dicholormethane at 31.2°C (After Filas, Hajdo & Eringen [1974])

Micropolar field theories turn out to contain just the right kind of degrees of freedom to explain and predict various physical phenomena relevant to liquid crystals. Several theoretical work employing micropolar theory or its special forms, through excellent experimental verifications, have demonstrated the relevance of the theory to the liquid crystal behavior cf. Lee and Eringen [1971a,b,c; 1973], Ericksen [1969], Leslie [1969], and Narasimhan and Eringen [1974]. Here we give two such results. In Fig. 9 are plotted the apparent viscosity and shear-rate dependence in a couette flow (based on micropolar theory) and compared with the experimental results of various investigators. The agreement with experiments is excellent. Note that Stokesian fluids display no such effect and therefore completely fails.

Fig. 10 displays the rotational motions of poly-γ-benzyl glutamate liquid crystals in a magnetic field. The measurement was made by nuclear magnetic resonance devices, Filas et al. [1974]. Note the unbelievably good agreement with the theoretical results. There are other calculations on wave propagations, orientational effects etc. Lee and Eringen [1971a,b,c].

Extensive applications of polar field theories exist in the areas of turbulence, blood flow, suspensions, composite materials, etc. For an extensive of these, we refer the reader to the review article by Ariman et al. [1973, 1974].

14. BALANCE LAWS OF NONLOCAL FIELD THEORIES

We learned that polar theories are capable of explaining and predicting a large class of microscopic phenomena in which the internal characteristic length is of the order of 10^{-5} cm or more and frequencies are of the order of magnitude 10^{13} cps or less. Beyond this region (in the range of atomic distances 10^{-8} cm for example) there exists much larger physical situations for which local theories (whether polar or not) fail to apply. In this range we have the causes of fatigue and fracture mechanism. Dislocations, impurities and surface physics come into play with critically affecting the material behavior. The approach through atomic lattice dynamics or quantum theory is formidable except for some ideal geometries and cohesive force laws which are often not realistic for most materials. In the remainder of this article, I would like to present a precis of nonlocal elasticity and discuss some of its applications. This field is rather new (not even 10 years old) and some major developments have taken place only recently. In fact, the approach which I will outline here is only in the publication stage Eringen [1976]*.

The nonlocal theory begins with the rejection of the *axiom of locality*. Therefore, the balance laws of nonlocal field theories contain the nonlocal effects (localization residuals $\hat{\phi}$ and $\hat{\Phi}$). In case of the nonlocal elasticity theory, the balance laws (6.9) to (6.13) are modified by these residuals so that we have

(i) *Mass*

$$\dot{\rho} + \rho v^k{}_{,k} = \hat{\rho} \qquad \text{in } V - \sigma$$
$$[\rho(v^k - \nu^k)]\, n_k = \hat{R} \qquad \text{on } \sigma \tag{14.1}$$

(ii) *Momentum*

$$t^{k\ell}{}_{,k} + \rho(f^\ell - \dot{v}^\ell) = \hat{\rho}\, v^\ell - \rho\hat{f}^\ell \qquad \text{in } V - \sigma$$
$$[\rho\, v^\ell (v^k - \nu^k) - t^{k\ell}]\, n_k = \hat{F}^\ell \qquad \text{on } \sigma \tag{14.2}$$

---

*For other earlier accounts See Eringen [1965, 1966, 1972c,d], Kunin [1966], Kröner [1966], Eringen and Edelen [1972].

(iii) *Moment of Momentum*

$$\varepsilon^{\ell mn} t_{mn} + \rho(\hat{\ell}^{\ell} - \varepsilon^{\ell mn} x_m \hat{f}_n) = 0 \qquad \text{in } V - \sigma$$

$$\hat{L}^{\ell} - \varepsilon^{\ell mn} x_m \hat{F}_n = 0 \qquad \text{on } \sigma \tag{14.3}$$

(iv) *Energy*

$$-\rho\dot{\varepsilon} + t^{k\ell} v_{\ell,k} + q^k{}_{,k} + \rho h - \hat{\rho}(\varepsilon - \frac{1}{2}\underset{\sim}{v}\cdot\underset{\sim}{v}) - \rho\hat{f}^k v_k + \rho\hat{h} = 0 \quad \text{on } V - \sigma$$

$$\underset{\sim}{[}\rho(\varepsilon + \frac{1}{2}\underset{\sim}{v}\cdot\underset{\sim}{v})(v^k - \nu^k) - t^{k\ell} v_{\ell} - q^k\underset{\sim}{]}\, n_k = \hat{H} \qquad \text{on } \sigma \tag{14.4}$$

(v) *Entropy*

$$\rho\dot{\eta} - (q^k/\theta)_{,k} - (\rho h/\theta) - \rho\hat{b} + \hat{\rho}\eta \geq 0 \qquad \text{in } V - \sigma$$

$$\underset{\sim}{[}\rho\eta(v^k - \nu^k) - (q^k/\theta)\underset{\sim}{]}\, n_k \geq \hat{B} \qquad \text{on } \sigma \tag{14.5}$$

Here ($\hat{\rho}$, $\hat{\underset{\sim}{f}}$, $\hat{\underset{\sim}{\ell}}$, $\hat{h}$, and $\hat{b}$) are the *nonlocal body residuals and* ($\hat{R}$, $\hat{F}$, $\underset{\sim}{L}$, $\hat{H}$, and $\hat{B}$) are the *nonlocal surface residuals* which are subject to

$$\int_{V-\sigma} (\hat{\rho}, \rho\hat{\underset{\sim}{f}}, \rho\,\hat{\underset{\sim}{\ell}}, \rho\,\hat{h}, \rho\hat{b})\, dv = 0$$

$$\int_{\sigma} (\hat{R}, \hat{\underset{\sim}{F}}, \hat{\underset{\sim}{L}}, \hat{H}, \hat{B})\, da = 0 \tag{14.6}$$

The physical significance of these residuals is clear from the equations in which they occur. For example, the mass residual, $\hat{\rho}$, represents the rate at which the mass is produced or destroyed at a point $\underset{\sim}{x}$ due to the other material points of the body occupying $V - \sigma$. Similarly $\hat{\underset{\sim}{f}}$ is the body force density at $\underset{\sim}{x}$ due to the interatomic attractions and repulsions of *all* other points of the body (e.g. gravitational, E-M attractions, etc.).

Similar meanings are obvious for other body and surface residuals. Clearly the determination of these residuals must be an integral part of the theory.

We introduce the following quantities for future use:

$$\psi \equiv \varepsilon - \theta\,\eta \quad , \quad Q^K = J\, q^k\, X^K_{\ ,k} \quad , \quad T^{K\ell} = J\, t^{k\ell}\, X^K_{\ ,k} \tag{14.7}$$

where $J \equiv \det (x^k_{\ ,K})$ is the Jacobian. Here ψ is the usual Helmholtz free energy function, $T^{K\ell}$ is called Piola-Kirchhoff stress tensor. By solving ε, $t^{k\ell}$ and q^k from these expressions and substituting into $(14.2)_1 - (14.3)_1$ and eliminating h between $(14.3)_1$ and $(14.4)_1$, we obtain

$$T^{K\ell}_{\ \ ,K} + \rho\, J(f^\ell - \dot{v}^\ell) = \hat{\rho}\, J\, v^\ell - \rho\, J\, \hat{f}^\ell \tag{14.8}$$

$$\varepsilon^{\ell mn}\, T^K_{\ n}\, x_{m,K} + \rho\, J(\hat{\ell}^\ell - \varepsilon^{\ell mn}\, x_m\, \hat{f}_n) = 0 \tag{14.9}$$

$$- \rho\, J\, \dot{\varepsilon} + T^K_{\ k}\, \dot{x}^k_{\ ,K} + Q^K_{\ ,K} + \rho\, Jh - \hat{\rho}\, J(\varepsilon - \frac{1}{2}\, \mathbf{v} \cdot \mathbf{v}) - \rho\, J\, \hat{f}^k\, \dot{x}_k + \rho\, J\hat{h} = 0 \tag{14.10}$$

$$- \frac{\rho J}{\theta}\, (\dot{\psi} + \dot{\theta}\eta) + \frac{1}{\theta}\, T^K_{\ k}\, \dot{x}^k_{\ ,K} + \frac{1}{\theta^2}\, Q^K\, \theta_{,K} - \frac{\hat{\rho} J}{\theta}\, (\psi - \frac{1}{2}\, \mathbf{v} \cdot \mathbf{v}) - \frac{\rho J}{\theta}\, \hat{\mathbf{f}} \cdot \dot{\mathbf{x}} + \frac{\rho J}{\theta}\, (\hat{h} - \theta\, \hat{b}) \geq 0 \tag{14.11}$$

These are the expressions of the balance laws in the reference frame.

15. NONLOCAL THERMODYNAMICS AND CONSTITUTIVE THEORY

The construction of the constitutive equations of nonlocal elastic solids begins by expressing ψ, $T^K_{\ k}$, Q^K, and η as functionals of the ordered *cause* set

$$\mathcal{F}' \equiv \{\mathbf{F}', \mathbf{F}'_{,L}\} \quad ; \quad \mathbf{F}' \equiv \{\mathbf{x}', \theta'\} \quad , \quad \mathbf{F}'_{,L} \equiv \{\mathbf{x}'_{,L}\, , \, \theta'_{,L}\} \tag{15.1}$$

where a prime indicates the dependence on a material $\underset{\sim}{X}' \varepsilon B$ and no prime on a reference point $\underset{\sim}{X}$, e.g.

$$\underset{\sim}{x}' = \hat{\underset{\sim}{x}}(\underset{\sim}{X}',t) \quad , \quad \underset{\sim}{x} = \hat{\underset{\sim}{x}}(\underset{\sim}{X},t) \tag{15.2}$$

Thus the free energy ψ is expressed as a functional in $\underset{\sim}{F}'$, $\underset{\sim}{X}'$ and a function of $\underset{\sim}{F}$ and $\underset{\sim}{X}$, i.e.

$$\psi(\underset{\sim}{X},t) = \hat{\psi}(\underset{\sim}{F}', \underset{\sim}{X}'; \underset{\sim}{F},\underset{\sim}{X}) = \hat{\psi}(\underset{\sim}{F}', \underset{\sim}{F}'_{,L}, \underset{\sim}{X}'; \underset{\sim}{F}, \underset{\sim}{F}_{,L}, \underset{\sim}{X}) \tag{15.3}$$

The smoothness requirements must now be placed on ψ. This requires that we consider a compact function space with topology. A Hilbert space is most convenient for this purpose. To this end, we define the inner product of two such functions $\underset{\sim}{F}_1'$ and $\underset{\sim}{F}_2'$ by

$$(\underset{\sim}{F}'_1, \underset{\sim}{F}'_2)_H \equiv \int_{V-\Sigma} H(|\underset{\sim}{X}' - \underset{\sim}{X}|)\, \underset{\sim}{F}_1(\underset{\sim}{X}') \cdot \underset{\sim}{F}_2(\underset{\sim}{X}')\, dV' \tag{15.4}$$

where $dV' \equiv dV(\underset{\sim}{X}')$ and

$$\underset{\sim}{F}_1(\underset{\sim}{X}') \cdot \underset{\sim}{F}_2(\underset{\sim}{X}') \equiv \underset{\sim}{x}'_1 \cdot \underset{\sim}{x}'_2 + \theta'_1\, \theta'_2 + \underset{\sim}{x}'_{1,K} \cdot \underset{\sim}{x}'_{2,K} + \theta'_{1,K}\, \theta'_{2,K} \tag{15.5}$$

The function $H(|\underset{\sim}{X}' - \underset{\sim}{X}|)$, called *alleviator*, is a positive decreasing function with $H(0) = 1$. The norm of $\underset{\sim}{F}'$ is now given by

$$||\underset{\sim}{F}'|| = (\underset{\sim}{F}',\underset{\sim}{F}')_H^{1/2} \tag{15.6}$$

The set of functions $\underset{\sim}{F}'$ now belongs to a Hilbert space, $\mathcal{H}$. If we assume that the argument functions in (15.3) are continuously differentiable and $\hat{\psi}$ is differentiable with respect to $\underset{\sim}{F}$, then the material time derivative of ψ exists and is given by

$$\dot{\psi}(\underset{\sim}{X},t) = \frac{\partial\hat{\psi}}{\partial\underset{\sim}{F}} \cdot \dot{\underset{\sim}{F}} + \int_{V-\Sigma} \frac{\delta\hat{\psi}}{\delta\underset{\sim}{F}'} \cdot \dot{\underset{\sim}{F}}'(\underset{\sim}{\Lambda})\, dV(\underset{\sim}{\Lambda}) \tag{15.7}$$

where $\delta\hat{\psi}/\delta F'$ is the Fréchet derivative of $\hat{\psi}$ with respect to F'. It is a functional of F', a function of F and a vector variable Λ. This representation (and hence $\delta\hat{\psi}/\delta\hat{F}'$) is unique since in a Hilbert space every linear functionals can be represented in the form of a scalar product.

For future convenience, we write (15.7) in the identical form

$$\rho J \dot{\psi} = [\rho J \frac{\partial\hat{\psi}}{\partial F} + \int_{V-\Sigma} (\rho J \frac{\delta\hat{\psi}}{\delta F'})^* dV(\Lambda)] \cdot \dot{F} + \mathcal{D} \tag{15.8}$$

where

$$\mathcal{D} \equiv \int_{V-\Sigma} [\rho J \frac{\delta\hat{\psi}}{\delta F'} \cdot \dot{F}(\Lambda) - (\rho J \frac{\delta\hat{\psi}}{\delta F'})^* \cdot \dot{F}(X)] dV(\Lambda) \tag{15.9}$$

Here and throughout an asterisk (*) indicates the interchange of X and Λ, e.g.,

$$[G(X, \Lambda)]^* = G(\Lambda, X) \tag{15.10}$$

For convenience, we also introduce the ordered sets

$$T = \{- \rho J \hat{f}_k , - \rho J \eta\} \quad , \qquad T^K \equiv \{T^K{}_k , 0 \} \tag{15.11}$$

Using these and (15.8) the entropy inequality (14.11) takes the form

$$\begin{aligned} &\frac{1}{\theta} (T - B) \cdot \dot{F} + \frac{1}{\theta} (T^L - B^L) \cdot \dot{F}_{,L} + \frac{Q^K}{\theta^2} \theta_{,K} - \frac{\mathcal{D}}{\theta} \\ &- \frac{\rho J}{\theta} (\hat{\psi} - \frac{1}{2} v \cdot v) + \frac{\rho J}{\theta} (\hat{h} - \theta\hat{b}) \geq 0 \end{aligned} \tag{15.12}$$

where

$$\begin{aligned} B &\equiv \rho J \frac{\partial\hat{\psi}}{\partial F} + \int_{V-\Sigma} (\rho J \frac{\delta\hat{\psi}}{\delta F'})^* dV(\Lambda) \\ B^L &\equiv \rho J \frac{\partial\hat{\psi}}{\partial F_{,L}} + \int_{V-\Sigma} (\rho J \frac{\delta\hat{\psi}}{\delta F'_{,L}})^* dV(\Lambda) \end{aligned} \tag{15.13}$$

Integration of (15.12) over V-Σ, after multiplication by θ > 0 gives

$$\int_{V-\Sigma} [(\underset{\sim}{T} - \underset{\sim}{B}) \cdot \dot{\underset{\sim}{F}} + (\underset{\sim}{T}^L - \underset{\sim}{B}^L) \cdot \dot{\underset{\sim}{F}}_{,L} + \frac{Q^K}{\theta} \theta_{,K} - \hat{\rho} J(\psi - \frac{1}{2} \underset{\sim}{v} \cdot \underset{\sim}{v}) - \rho J \theta \hat{b}] \, dV(\underset{\sim}{X}) \geq 0 \tag{15.14}$$

since on account of the skew-symmetry of the integrand of $\mathcal{D}$ in $\underset{\sim}{X}$ and $\underset{\sim}{\Lambda}$ and because of the conditions (13.6), the integrals of $\mathcal{D}$ and $\rho J\hat{h}$ over V-Σ vanish.

Clearly when two spatial reference frames differ by a rigid body motion, there should be no change in the physical characteristics of the body. Thus, we posit that *the constitutive functionals and residuals must be form-invariant* under "Galilean transformations" of the spatial frame of reference, expressed by

$$\begin{aligned} &\bar{\underset{\sim}{x}}' = \underset{\sim}{Q}\, \underset{\sim}{x}' + \underset{\sim}{V} t + \underset{\sim}{b} \quad , \quad &&\bar{\underset{\sim}{x}} = \underset{\sim}{Q}\, \underset{\sim}{x} + \underset{\sim}{V} t + \underset{\sim}{b} \\ &\underset{\sim}{Q}\, \underset{\sim}{Q}^T = \underset{\sim}{Q}^T \underset{\sim}{Q} = \underset{\sim}{I} \quad , \quad &&\det \underset{\sim}{Q} = \pm 1 \end{aligned} \tag{15.15}$$

where $\underset{\sim}{Q}$ is a *constant* orthogonal transformation and $\underset{\sim}{V}$ and $\underset{\sim}{b}$ are *constant* vectors. This postulate is much milder than the axion of objectivity used for the local theories. Thus $\hat{\rho}$, $\hat{h}$, $\hat{b}$, and the coefficients of $\underset{\sim}{F}$, $\underset{\sim}{F}_{,L}$ Q^K, and ψ cannot depend on $\underset{\sim}{v}$. Because of the conditions $(14.6)_1$ on $\hat{\rho}$ the volume integral of $\hat{\rho} J \underset{\sim}{v} \cdot \underset{\sim}{v}$ cannot be made non-negative unless

$$\hat{\rho} = 0 \tag{15.16}$$

and therefore $\rho J = \rho_o$. Using the identity

$$(\underset{\sim}{T}^L - \underset{\sim}{B}^L) \cdot \underset{\sim}{F}_{,L} = [(\underset{\sim}{T}^L - \underset{\sim}{B}^L) \cdot \dot{\underset{\sim}{F}}]_{,L} - (\underset{\sim}{T}^L - \underset{\sim}{B}^L)_{,L} \cdot \dot{\underset{\sim}{F}}$$

and the Green-Gauss theorem one part of the second term in (15.14) can be converted to a surface integral. Thus

$$\int_{V-\Sigma} [(\underset{\sim}{T} - \underset{\sim}{B} - \underset{\sim}{T}^L{}_{,L} + \underset{\sim}{B}^L{}_{,L}) \cdot \dot{\underset{\sim}{F}} \, dV(\underset{\sim}{\Lambda}) + \int_{S-\Sigma} (\underset{\sim}{T}^L - \underset{\sim}{B}^L) \cdot \dot{\underset{\sim}{F}} \, dA_L(\underset{\sim}{X})$$

$$- \int_{\Sigma} [(\underset{\sim}{T}^L - \underset{\sim}{B}^L) \cdot \dot{\underset{\sim}{F}}] \, dA_L(\underset{\sim}{X}) + \int_{V-\Sigma} (\frac{Q^K}{\theta} \theta_{,K} - \rho_o \theta \hat{b}) \, dV(\underset{\sim}{X}) \geq 0$$

This inequality is linear in $\dot{\underset{\sim}{F}}$. Thus for arbitrary variations of this function throughout V - Σ, S - Σ and Σ it cannot be maintained in one sign unless

$$\begin{aligned} &\underset{\sim}{T} - \underset{\sim}{B} - \underset{\sim}{T}^{L}{}_{,L} + \underset{\sim}{B}^{L}{}_{,L} = \underset{\sim}{0} && \text{in } V - \Sigma \\ &(\underset{\sim}{T}^{L} - \underset{\sim}{B}^{L})\, N_L = \underset{\sim}{0} && \text{on } S - \Sigma \qquad (15.17) \\ &[\underset{\sim}{T}^{L} - \underset{\sim}{B}^{L}]\, N_L = 0 && \text{on } \Sigma \end{aligned}$$

where N_L denotes the positive unit normal of S - Σ and Σ and

$$\int_{V-\Sigma} \left(\frac{Q^K}{\theta}\,\theta_{,K} - \rho_o\,\theta\,\hat{b}\right) dV(\underset{\sim}{X}) \geq 0 \qquad (15.18)$$

we have therefore proved

Theorem. The constitutive equations and the residuals of the nonlocal thermoelastic solids do not violate the global entropy inequality, if and only if they satisfy (15.17) and do not violate (15.18).

Carrying (15.1) and (15.11) into (15.17) we obtain

$$\begin{aligned} &\rho_o\,\hat{f}_k = \rho_o\,\check{f}_k + \check{T}^K{}_{k,K} - T^K{}_{k,K} \\ &\rho_o\eta = \rho_o\check{\eta} + (\check{Q}^K/\theta)_{,K} && \text{in } V - \Sigma \\ &(T^K{}_k - \check{T}^K{}_k)\, N_K = 0 \quad , \quad 0 = \frac{\check{Q}^K}{\theta} N_K && \text{on } S - \Sigma \qquad (15.19) \\ &[\underset{\sim}{T}^K{}_k - \check{T}^K{}_{\underset{\sim}{k}}]\, N_K = 0 && \text{on } \Sigma \end{aligned}$$

where we define

$$\rho_o\,\check{f}_k \equiv -\rho_o\,\frac{\partial\hat{\psi}}{\partial x^k} - \int_{V-\Sigma} \left(\rho_o\,\frac{\delta\hat{\psi}}{\delta x'^k}\right)^* dV(\underset{\sim}{\Lambda})$$

$$\check{T}^K{}_k \equiv \rho_o\,\frac{\partial\hat{\psi}}{\partial x^k{}_{,K}} + \int_{V-\Sigma} \left(\rho_o\,\frac{\delta\hat{\psi}}{\delta x'^k{}_{,K}}\right)^* dV(\underset{\sim}{\Lambda})$$

$$\rho_o \check{\eta} \equiv - \rho_o \frac{\partial \hat{\psi}}{\partial \theta} - \int_{V-\Sigma} (\rho_o \frac{\delta \hat{\psi}}{\delta \theta'})^* \, dV(\underset{\sim}{\Lambda})$$

$$\frac{\check{Q}^K}{\theta} \equiv \rho_o \frac{\partial \hat{\psi}}{\partial \theta_{,K}} + \int_{V-\Sigma} (\rho_o \frac{\delta \hat{\psi}}{\delta \theta'_{,K}})^* \, dV(\underset{\sim}{\Lambda}) \tag{15.20}$$

It is interesting to note that $T^K{}_k$ is determined only on the surface of the body $S - \Sigma$ as against the local theory where it is determined at all points of the body. If we recall that the stress is a concept always introduced by the resolution of the surface traction, the present result acquires a deeper meaning. An even deeper significance of these results is revealed if we introduce (15.19) into the balance laws and entropy inequality and use (15.16)

$$\check{T}^K{}_{k,K} + \rho_o (f_k - \dot{v}_k) + \rho_o \check{f}_k = 0$$

$$\varepsilon^{\ell mn} x_{m,K} \check{T}^K{}_n + \rho_o (\check{\ell}^\ell - \varepsilon^{\ell mn} x_m \check{f}_n) = 0 \tag{15.21}$$

$$- \rho_o \theta \dot{\eta} + Q^K{}_{,K} + \rho_o (h + \hat{h}) = 0$$

$$\frac{1}{\theta} Q^K \theta_{,K} + \rho_o (h - \theta \hat{b}) \geq 0$$

where we defined

$$\rho_o \check{\ell}^\ell \equiv \rho_o \check{\ell}^\ell - \varepsilon^{\ell mn} [(\check{T}^K{}_n - T^K{}_n) x_m]_{,K}$$

$$\rho_o \check{h} \equiv \rho_o \hat{h} - \mathcal{D} + [(T^K{}_k - \check{T}^K{}_k) \dot{x}_k - \frac{Q^K}{\theta} \dot{\theta}]_{,K} \tag{15.22}$$

We observe that (15.21) are identical in form to (14.8) - (14.11) (with $\hat{\rho} = 0$ as found). Therefore, if we consider $\check{T}^K{}_k$ and $\check{f}^k$, $\check{\ell}^k$ and $\check{h}$ as the stress and new residuals the field theory so obtained will be identical to the nonlocal field theory we intended to construct, provided of course, the new residuals satisfy the conditions

$$\int_{V-\Sigma} \rho_o(\check{\underset{\sim}{f}}, \check{\underset{\sim}{\ell}}, \check{h})\, dV(\underset{\sim}{X}) = 0 \tag{15.23}$$

The fact that (15.23) is satisfied when $\hat{f}$, $\hat{\underset{\sim}{\ell}}$ and $\hat{h}$ satisfy (14.6) is clear from the conditions (15.14). Conversely (15.23) implies (14.6).

We now proceed to show that (15.23) is satisfied. To this end we impose Galilean invariance on $\hat{\psi}$. For $\underset{\sim}{Q} = \underset{\sim}{I}$, $\underset{\sim}{V} = \underset{\sim}{0}$ and $\underset{\sim}{b} = \underset{\sim}{x}$, it follows that $\hat{\psi}$ can depend on $\underset{\sim}{x}$ only through $\underset{\sim}{x}' - \underset{\sim}{x}$. Thus

$$\frac{\partial\hat{\psi}}{\partial x^k} = - \int_{V-\Sigma} \frac{\delta\hat{\psi}}{\delta(x'^k - x^k)}\, dV(\underset{\sim}{\Lambda}) = - \int_{V-\Sigma} \frac{\delta\hat{\psi}}{\delta x'^k}\, dV(\underset{\sim}{\Lambda})$$

Substituting this into $(15.20)_1$ we will have

$$\rho_o\, \check{f}^k = \int_{V-\Sigma} [\rho_o \frac{\delta\hat{\psi}}{\delta x'^k} - (\rho_o \frac{\delta\hat{\psi}}{\delta x'^k})^*]\, dV(\underset{\sim}{\Lambda}) \tag{15.24}$$

The volume integral of this over $V - \Sigma$ vanishes so that $(15.23)_1$ is satisfied.

Galilean invariance under rigid rotations $\{\underset{\sim}{Q}\}$ can be used to show that cf., Eringen [1975]

$$\begin{aligned} \rho_o\, \check{\ell}^k = \varepsilon^{k\ell m} \int_{V-\Sigma} & [\rho_o\, x_\ell(\underset{\sim}{\Lambda}) \frac{\delta\hat{\psi}}{\delta x'^m} - x_\ell(\underset{\sim}{X})(\rho_o \frac{\delta\hat{\psi}}{\delta x'^m})^* \\ & + \rho_o\, x^\ell{}_{,K}(\underset{\sim}{\Lambda}) \frac{\delta\hat{\psi}}{\delta x'^m{}_{,K}} - x^\ell{}_{,K}(\underset{\sim}{X})(\rho_o \frac{\delta\hat{\psi}}{\delta x'^m{}_{,K}})^*]\, dV(\underset{\sim}{\Lambda}) \end{aligned} \tag{15.25}$$

Since the integrand is skew-symmetric this too integrates to zero, over $V - \Sigma$.

Similarly Galilean invariance for $\rho_o h$ and the energy equation $(15.21)_3$ gives

$$(T^K{}_k - \check{T}^K{}_k)_{,K} = \int_{V-\Sigma} [\rho_o \frac{\delta\hat{\psi}}{\delta x'^k} - (\rho_o \frac{\delta\hat{\psi}}{\delta x'^k})^*]\, dV(\underset{\sim}{\Lambda}) = \rho_o(\check{f}_k - \hat{f}_k) \tag{15.26}$$

which serves merely to determine $\hat{f}_K$ that is no longer needed in the set (15.21).

The invariance of $\hat{\psi}$ under Galilean transformations of the spatial frame of reference also shows that $\hat{\psi}$ must be a functional of

$$\kappa'_K \equiv (x'_k - x_k)\, x^k{}_{,K} \quad , \quad B'_{KL} \equiv x'^k{}_{,K}\, x_{k,L} - \delta_{KL} \tag{15.27}$$

and a function of C_{KL}, i.e.

$$\rho_o\, \psi(\underset{\sim}{X},t) = \Sigma(\kappa'_K, B'_{KL}, \theta', \theta'_{,K}, \underset{\sim}{X}' \;;\; C_{KL}, \theta, \theta_{,K}, \underset{\sim}{X}) \tag{15.28}$$

With these (15.20) and $(15.21)_2$ gives

$$\check{T}^K{}_k = 2\frac{\partial\Sigma}{\partial C_{KL}} x_{k,L} + \int_{V-\Sigma} \{\frac{\delta\Sigma}{\delta\kappa'_K}[x_k(\underset{\sim}{\Lambda}) - x_k(\underset{\sim}{X})] + [\frac{\delta\Sigma}{\delta B'_{LK}} + (\frac{\delta\Sigma}{\delta B'_{KL}})^*]x_{k,L}(\underset{\sim}{\Lambda})\}dV(\underset{\sim}{\Lambda})$$

$$\rho_o\, \check{f}^k = \int_{V-\Sigma} [\frac{\delta\Sigma}{\delta\kappa'_K} x^k{}_{,K}(\underset{\sim}{X}) - (\frac{\delta\Sigma}{\delta\kappa'_K})^* x^k{}_{,K}(\underset{\sim}{\Lambda})]\, dv(\underset{\sim}{\Lambda})$$

$$\rho_o\, \check{\ell}^k = \varepsilon^{k\ell m}\int_{V-\Sigma} [(\frac{\delta\Sigma}{\delta\kappa'_K})^* x_{\ell,K}(\underset{\sim}{\Lambda})\, x_m(\underset{\sim}{X}) - \frac{\delta\Sigma}{\delta\kappa'_K} x_{\ell,K}(\underset{\sim}{X})\, x_m(\underset{\sim}{\Lambda}) \tag{15.29}$$

$$+ (\frac{\delta\Sigma}{\delta B'_{KL}})^* x_{\ell,K}(\underset{\sim}{\Lambda})\, x_{m,K}(\underset{\sim}{X}) - \frac{\delta\Sigma}{\delta B'_{LK}} x_{\ell,L}(\underset{\sim}{X})\, x_{m,K}(\underset{\sim}{\Lambda})]\, dV(\underset{\sim}{\Lambda})$$

$$\rho_o \eta = -\frac{\partial\Sigma}{\partial\theta} - \int_{V-\Sigma} (\frac{\delta\Sigma}{\delta\theta'})^*\, dV(\underset{\sim}{\Lambda})$$

For Q^K we need to write (as in classical theory) a constitutive equation. For non-heat conducting solids, no need arises for this or for the determination of $\check{h}$ and $\check{b}$. For heat conducting materials, however, one needs constitutive equations for these quantities.

The global entropy inequality (15.18) in the case of isothermal motions reduces to the classical form

$$\int_{V-\Sigma} \frac{Q^K}{\theta} dV(\underset{\sim}{X}) \geq 0 \tag{15.30}$$

since θ = const.

The exact constitutive equations and residuals determined by (15.29) can be used to derive various order theories. In the following section, we give the linear theory.

16. LINEAR THEORY OF NONLOCAL ELASTICITY

For simplicity, here we obtain the constitutive equations of linear, nonlocal, isothermal elastic solids. For this, the free energy is assumed to be quadratic in the strain measures, however being linear in the nonlocal measures. From (14.3), it is clear that the stress tensor in the nonlocal theory may not be symmetric. If we impose the condition of symmetry for the stress tensor, then we must have

$$\hat{\underset{\sim}{\ell}} = \underset{\sim}{x} \times \hat{\underset{\sim}{f}} \tag{16.1}$$

Using (15.29) these conditions can be satisfied if (cf. Eringen [1975])

$$\Sigma = \Sigma\ (r'^2,\ E'_{KL}\ ;\ E_{KL},\ \theta,\ \underset{\sim}{X}) \tag{16.2}$$

where

$$r'^2 \equiv |\underset{\sim}{x}' - \underset{\sim}{x}|^2 \quad , \quad 2E'_{KL} = x'^k{}_{,K}\, x'_{k,L} - \delta_{KL} \tag{16.3}$$

The stress tensor $t^{k\ell}$ and residuals $\check{f}^k$ and $\check{\ell}^k$ can now be derived by using (15.29) and $(14.7)_2$ where $\check{T}^K{}_k$ is used in place of $T^K{}_K$.

$$t^{k\ell} = \frac{\rho}{\rho_o} [\frac{\partial\Sigma}{\partial E_{KL}} + \int_{V-\Sigma} (\frac{\delta\Sigma}{\delta E'_{KL}})^* dV'] \, x^{(k}{}_{,K}\, x^{\ell)}{}_{,L}$$

$$+ \frac{\rho}{\rho_o} \int_{V-\Sigma} [\frac{\delta\Sigma}{\delta E'_{KL}} \, x'^{(k}{}_{,K}\, x'^{\ell)}{}_{,L} + 2\frac{\delta\Sigma}{\delta r'^2} (x'^k - x^k)(x'^\ell - x^\ell)]\, dV'$$

$$\rho_o f^k = 2 \int_{V-\Sigma} [\frac{\delta\Sigma}{\delta r'^2} + (\frac{\delta\Sigma}{\delta r'^2})^*] (x'^k - x^k)\, dV'$$

$$\rho_o \ell^k = 2 \varepsilon^{k\ell m} \int_{V-\Sigma} [\frac{\delta\Sigma}{\delta r'^2} + (\frac{\delta\Sigma}{\delta r'^2})^*] x_\ell x'_m\, dV'$$

where parenthesis enclosing indices indicate symmetrization, e.g.

$$x^{(k}{}_{,K}\, x^{\ell)}{}_{,L} \equiv \frac{1}{2} (x^k{}_{,K}\, x^\ell{}_{,L} + x^\ell{}_{,K}\, x^k{}_{,L})$$

These are the final forms for the nonlinear theory in the case of the symmetric stress tensor. For the construction of the linear theory, we write

$$\Sigma = \Sigma_o (\underset{\sim}{E}, \theta, \underset{\sim}{X}) + \int_{V-\Sigma} (\Sigma_o' + \Sigma_1'^{MN} E'_{MN})\, dV' \tag{16.5}$$

where

$$\Sigma_o(\underset{\sim}{E}, \theta, \underset{\sim}{X}) = \Sigma_{oo} + \Sigma_{01}^{KL} E_{KL} + \frac{1}{2} \Sigma_{02}^{KLMN} E_{KL} E_{MN}$$

$$\Sigma_o'^2 (r'^2, \underset{\sim}{X}' ; \underset{\sim}{E}, \theta, \underset{\sim}{X}) = \Sigma_{oo}' + \Sigma_{01}'^{KL} E_{KL} + \frac{1}{2} \Sigma_{02}'^{KLMN} E_{KL} E_{MN} \tag{16.6}$$

$$\Sigma_1'^{MN} (r'^2, \underset{\sim}{X}' ; \underset{\sim}{E}, \theta, \underset{\sim}{X}) = \Sigma_{10}'^{MN} + \Sigma_{11}'^{MNKL} E_{KL}$$

With these and the usual linearization process, after some lengthy manipulations, we arrive at

$$t_{k\ell} = \lambda e^r{}_r \delta_{k\ell} + 2\mu e_{k\ell} + \int_{V-\sigma} (\lambda' e'^r{}_r \delta_{k\ell} + 2\mu' e'_{k\ell})\, dv'$$

$$\check{f}^k = \check{\ell}^k = 0 \tag{16.7}$$

$$\Sigma = \alpha_o + \frac{\lambda}{2} (e^k{}_k)^2 + \mu e_{k\ell} e^{k\ell} + \frac{1}{2} \int_{V-\sigma} (\lambda' e_{kk} e'^{\ell\ell} + 2\mu' e_{k\ell} e'^{k\ell})\, dv'$$

where

$$\lambda' = \lambda'(|\underset{\sim}{x}' - \underset{\sim}{x}|, \theta) \quad , \quad \mu' = \mu'(|\underset{\sim}{x}' - \underset{\sim}{x}|, \theta) \tag{16.8}$$

are the *nonlocal elastic moduli* for the homogeneous and isotropic solids. We note that in the derivation of (16.7) we have used the isotropic expressions of $\Sigma_{\alpha\beta}^{KL\cdot\cdot}$ and made passage to the spatial expressions of the material tensors, e.g.

$$e_{k\ell} = E_{KL}\,\delta^K{}_k\,\delta^L{}_\ell = \frac{1}{2}(u_{k,\ell} + u_{\ell,k})$$

Also, the terms multiplied by $X'^K - X^K$ have been dropped. These terms, interesting as they are,[1] involve interatomic distances so that they are negligibly small. For example, the isotropic forms of Σ'^K and Σ'^{KL} are

$$\Sigma'^K = \Sigma_o'(X'^K - X^K),$$

$$\Sigma'^{KL} = \Sigma_1'\,\delta^{KL} + \Sigma_2'(X'^K - X^K)(X'^L - X^L)$$

where Σ_α' depend on $|\underset{\sim}{X} - \underset{\sim}{X}|$ and θ. Thus Σ_o' and Σ_2' have been dropped because of ignoring Σ_o', the body force and couple residuals vanish. This approximation is not generally permissible when the gravitational long range forces are important. In that case $\check{\underset{\sim}{f}}$ and $\check{\underset{\sim}{\ell}}$ do not vanish, see Eringen [1976].

Field Equations. The field equations follow from (16.7) and $(14.2)_1$.

$$(\lambda + 2\mu)\,\underset{\sim}{\nabla}\underset{\sim}{\nabla}\cdot\underset{\sim}{u} - \mu\,\underset{\sim}{\nabla}\times\underset{\sim}{\nabla}\times\underset{\sim}{u} + \int_{V-\sigma}[(\lambda' + 2\mu')\,\underset{\sim}{\nabla}'\times\underset{\sim}{\nabla}'\times\underset{\sim}{u}' - \mu'\underset{\sim}{\nabla}'\times\underset{\sim}{\nabla}'\times\underset{\sim}{u}']dv'$$

$$- \int_{S-\sigma}\underset{\sim}{T}'^k\,da'_k + \int_\sigma[\underset{\sim}{T}'^k]\,da'_k + \rho(\underset{\sim}{f} - \ddot{\underset{\sim}{u}}) = \underset{\sim}{0} \tag{16.9}$$

---

[1]Interestingly enough these terms can be used to deal with material symmetries in the atomic and molecular scales. For example, a solid with cubic symmetry in the atomic scale (e.g. alcaly Halides) could be isotropic in the global scale.

where

$$\underset{\sim}{T}'^{k} = T'^{k\ell} \underset{\sim}{i}_{\ell} \quad , \qquad T'_{k\ell} \equiv \lambda' e'^{r}{}_{r} \delta_{k\ell} + 2\mu' e'_{k\ell} \tag{16.10}$$

are the surface tractions and surface stress densities. The presence of these surface integrals in (16.9) indicate that the *nonlocal theory include the surface physics*. Thus, for example, the surface tension, surface shear, surface energy (all of which are missing in the local theories) are all integral parts of the nonlocal theory.

Alternative forms of (16.7) and (16.8) are

$$t_{k\ell} = \int_{V-\sigma} (\hat{\lambda} e'^{r}{}_{r} \delta_{k\ell} + 2\hat{\mu} e_{k\ell}) \, dv' \tag{16.11}$$

$$\int_{V-\sigma} [(\hat{\lambda} + 2\hat{\mu}) \underset{\sim}{\nabla}'\underset{\sim}{\nabla}' \cdot \underset{\sim}{u}' - \hat{\mu} \underset{\sim}{\nabla}' \times \underset{\sim}{\nabla}' \times \underset{\sim}{u}'] \, dv' + \rho(\underset{\sim}{f} - \ddot{\underset{\sim}{u}}) - \int_{S-\sigma} \underset{\sim}{T}'^{k} \, da'_{k} + \int_{\sigma} [\underset{\sim}{T}'^{k}] \, da'_{k} = \underset{\sim}{0} \tag{16.12}$$

where λ and μ are incorporated into $\hat{\lambda}$ and $\hat{\mu}$. This is permissible provided we understand that $\hat{\lambda}$ and $\hat{\mu}$ become Dirac delta measures in the limit when the nonlocal continuum becomes local (cf. Section 17).

17. DISPERSION OF PLANE WAVES

The one-dimensional form of the field equations (16.8) without surface and body forces is obtained by setting

$$u_1 = u(x,t) \quad , \quad u_2 = u_3 = 0 \quad , \quad \underset{\sim}{f} = \underset{\sim}{0} \quad , \quad (x_1,x_2,x_3) = (x,y,z) \tag{17.1}$$

Assuming that the body is of infinite extend, (16.1) can be written as

$$\int_{-\infty}^{\infty} k'(|x' - x|) \frac{\partial^2 u'}{\partial x'^2} dx' = \frac{1}{c_o^2} \frac{\partial^2 u}{\partial t^2} \tag{17.2}$$

where

$$k' \equiv \int_{-\infty}^{\infty} \int_{-\infty}^{\infty} \frac{\hat{\lambda} + 2\mu}{\lambda + 2\mu} dy' \, dz' \quad , \quad c_o^2 \equiv \frac{\lambda + 2\mu}{\rho} \tag{17.3}$$

We consider Fourier transform solution of (17.2), i.e.

$$\bar{u}(\xi,\omega) \equiv \int_{-\infty}^{\infty} \int_{-\infty}^{\infty} u(x,t) \exp (i\xi x + i\omega t) \, dx \, dt \tag{17.4}$$

Applying the Fourier transform to (17.2) we obtain

$$\omega^2/c_o{}^2 = \bar{k}'(\xi) \, \xi^2 \tag{17.4}$$

This shows that the phase velocity depends on the wave number and hence *the waves in nonlocal bodies are dispersive.* From the lattice dynamics for the one-dimensional Born-von Kármán model, we have

$$\bar{k}'(\xi) = (4/\xi^2 a^2) \sin^2 (\xi\varepsilon/2) \tag{17.5}$$

where a is the atomic distance in a chain of equidistant atoms attached by linear springs. If (17.4) is made to coincide with (17.5) in the entire Brillouin zone $|\xi a| \leq \pi$ then we must have

$$\bar{k}'(\xi) = (4/\xi^2 a^2) \sin^2 (\xi a/2) \tag{17.6}$$

Inverse Fourier transform of this gives

$$k(|x|) = \begin{cases} \frac{1}{a}\left(1 - \frac{|x|}{a}\right) & , \quad |x| \leq a \\ 0 & , \quad |x| > a \end{cases} \tag{17.7}$$

This simple and elegant result indicates that if the nonlocal moduli is taken as in (17.7), then the nonlocal continuum theory will predict identical results to the atomic lattice dynamics. In Fig. 11 , there is displayed $k(|x|)$ versus $|x|/a$. Note that the area under $k(|x|)$ is unity. Hence as $a \to 0$, (17.7) gives a dirac delta measure so that *when the atomic distance appraoches to zero the nonlocal theory goes to the classical theory.*

Further, since the analysis is the same for the shear waves for which μ and μ' replace $\lambda + 2\mu$ and $\lambda' + 2\mu'$, it is suggested that

$$\left(\frac{\hat{\lambda}}{\lambda}, \frac{\hat{\mu}}{\mu}\right) = \begin{cases} \frac{1}{a}\left(1 - \frac{|\underset{\sim}{x}|}{a}\right) & , \quad |\underset{\sim}{x}| \leq a \\ 0 & \quad |\underset{\sim}{x}| \geq a \end{cases} \tag{17.8}$$

is a good candidate for the nonlocal moduli. Of course other possibilities exist and in fact the experimental results on phonon dispersion can be employed to obtain physically more realistic forms.

By use of this form, I gave a solution of the problem for the elastic surface waves, Eringen [1973a, 1974]. Without any curve fitting the resulting dispersion curves is shown in Fig. 12. Note the unbelievable coincidence with the results of the atomic lattice theory.

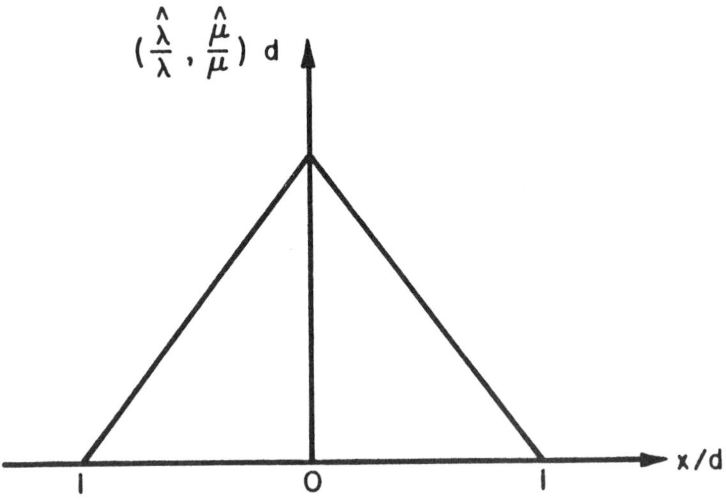

Fig. 11 Nonlocal Elastic Moduli

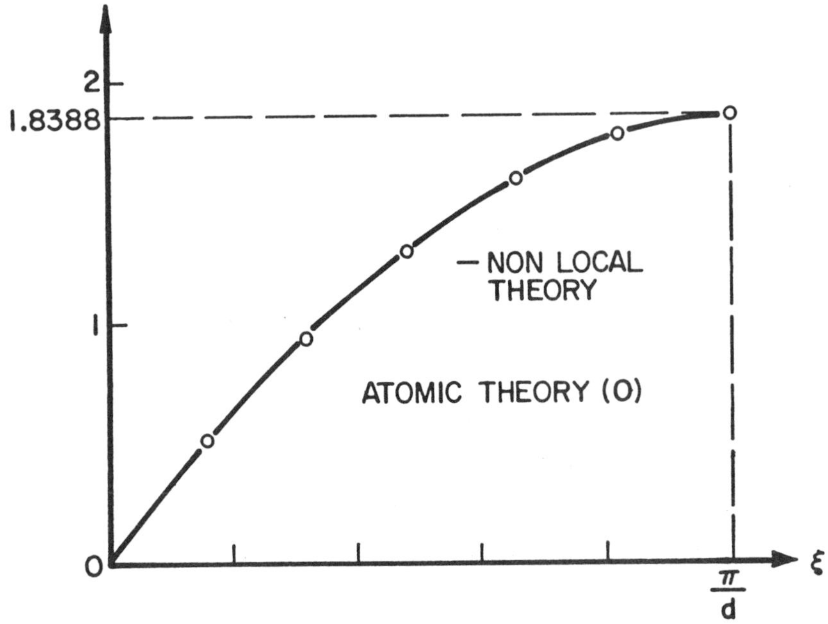

Fig. 12 Dispersion of Surface Waves

18. FRACTURE MECHANICS

A crucial problem in fracture mechanics is the fracture criterion which is supposed to predict the initiation of crack. Classical elasticity solution of the problem of a plate, with a line crack, subject to uniform tension, perpendicular to the line of crack, at infinity predicts a hoop stress distribution along the line of crack which is infinite at the crack tip. The stress singularity is of the order 1/ r where r is the distance from the crack tip. Stress singularities occur also in nonlinear elasticity and in micropolar theory. Because of these stress singularities all fracture criteria hitherto introduced in fracture mechanics had to circumvent the natural concepts based on critical stresses (e.g. yield stress). With the celebrated work of Griffith all fracture criteria avoid the stress but rather employ such concepts as stress intensity factor, fracture toughness or other similar concepts (e.g. J-integral). These fracture criteria and their various modifications always required experimental verifications in each case. We shall now show briefly that this situation does not arise in the nonlocal theory and the long-standing, unresolved question can now be answered.

For simplicity, we assume that the intermolecular long range effects are in the direction of the line of crack only and neglect u_2-component of the displacement field in that direction. With this and dropping the surface integrals, (16.1) for the two-dimensional case, reduces to

$$\int_{-\infty}^{\infty} [(\hat{\lambda} + 2\hat{\mu}) \frac{\partial^2 v(x',y)}{\partial y^2} + \hat{\mu} \frac{\partial^2 v(x',y)}{\partial x'^2}] \, dx' = 0 \qquad (18.1)$$

where $v \equiv u_2$.

The boundary conditions for a pressurized crack are

$$\begin{aligned} &v = 0 && \text{when} && y = \infty && \\ &v = 0 && \text{when} && y = 0 , && |x| \geq \ell \\ &t_{yy} = -\sigma_o && \text{when} && y = 0 , && |x| < \ell \end{aligned} \qquad (18.2)$$

To the solution so obtained by superimposing a simple solution for the plate with no crack and $t_{yy} = \sigma_o$, we obtain the solution of the crack problem (Fig. 13). Eringen and Kim [1974] found a solution of the integro-differential equation (18.1) subject to the *mixed* boundary conditions (18.2) by solving the ensuing dual integral equations. We only give the most crucial result here. This is

$$\frac{\sigma}{\sigma_o} \simeq \frac{2}{3} \left(\frac{2\ell}{a}\right)^{1/2} \qquad (18.3)$$

where $\sigma \equiv t_{yy}(\ell,0)$ is the hoop stress at the crack tip and σ_o is the tensile stress at infinity.

One can appreciate the importance of this result by the following observations:

(a) The stress singularity for $a \to 0$ is the same as in classical elasticity.

(b) Since the nonlocal elasticity gives a finite stress at the crack tip, we are able to propose a natural fracture criterion based on the maximum stress:

Fracture Criterion. When the tensile stress at crack tip reaches a critical value σ_c, the crack begins to expand.

Note how simple, natural and physically resonable this is. The question may be raised what is σ_c. Clearly, we are dealing with forces in the atomic scale. Thus σ_c must be the *cohesive stress*. For steel this is about $\sigma_c = E/5.5$ where E is the Youngs modulus. If we use a = 3Å, we find that the fracture toughness K_{Ic} (as defined by ASTM) is

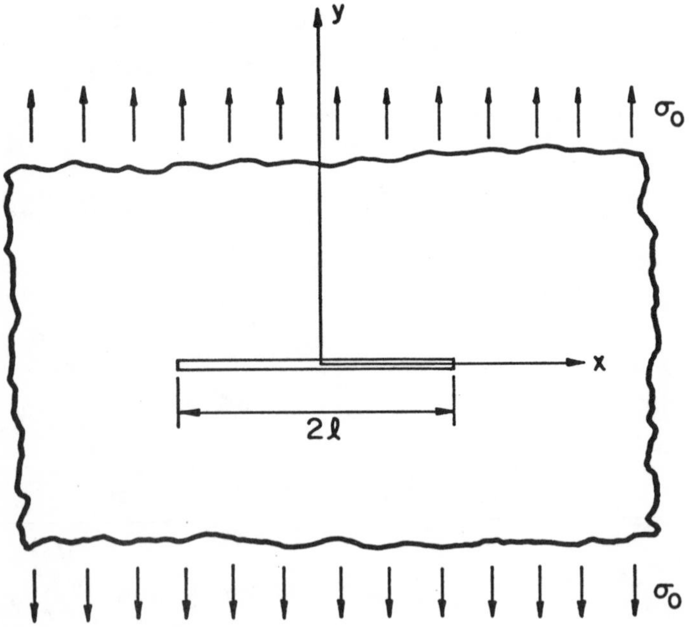

Fig. 13 Crack Problem

$$K_{Ic} = 41.9\ \sigma_o \sqrt{\ell} = 26{,}300 \text{ psi } \sqrt{\text{in}}$$

where (18.3) is used with $\sigma = \sigma_c$. This result is in fair agreement with the experimental results $K_{Ic} = 35{,}000$ psi$\sqrt{\text{in}}$ for steel at about -100°F [cf. ASTM, STP 514, 1971, p. 166]. Similarly for glass, calculations give $K_{Ic} = 7{,}500$ psi $\sqrt{\text{in}}$ which compares well with the experimental value 10,000 psi$\sqrt{\text{in}}$, as obtained by Griffith experimentally.

(c) If in (18.3) we take $\sigma = \sigma_c$ = the molecular cohesive stress, we obtain

$$\sigma^2 \ell = C \quad , \quad C \equiv \frac{9}{8} \sigma_c^{\,2} d \tag{18.4}$$

This is exactly the *Griffith criterion* for static fracture except that here Griffith constant C is fully specified.

These examples should suffice to illustrate great power and potential of the nonlocal field theories.

19. PROSPECT

The account presented here is only a birds-eye view of the recent developments in continuum physics. Raison d'etre for this account is two-fold: (a) to draw attention of the research workers in other fields of physical sciences to the fundamental ideas and great potential of newly opened fields in continuum physics and (b) to emphasize the fact that physical sciences, irrespective of different models, must strive towards unity since only then may we begin to understand and control the physical phenomena.

I have tried to indicate, with few examples, that indeed the land between atomic and the macroscopic scales is extremely rich in interesting physical phenomena and in dealing with real materials we are forced to investigate these barren regions of research: the escape to the two ideal extremes will never provide the answers to the real problems that reside in this region.

I have not touched many important topics concerned with the materials that possess internal structure. Polarized media theories, nonlocal electromagnetism, composites, polymers, interacting media, plasma, chemically active media, mixtures, dislocation, surface physics, etc. are all ignored. While much new and comprehensive studies in these fields exist, new research is needed for a definitive organization of some of these fields. Accounts on some of these subjects are to be found in continuum physic series I edited (cf. [1976] Vols. III and IV). For microelectromagnetism see Eringen and Kafadar [1970] for relativistic continua Grot and Eringen [1966], Eringen [1970c], for nonlocal electromagnetism Eringen [1973b], for nonlocal micromorphic solids and fluids Eringen [1973c,d].

REFERENCES

Ariman, T., Sylvester, N. D. & Turk, M. A. [1973] Microcontinuum Fluid Mechanics - A Review *Int. J. Engng. Sci.* 11, 905.

Ariman, T., Turk, M. A. & Sylvester, N. D. [1974] Applications of Micro-continuum Fluid Mechanics *Int. J. Engng. Sci.* 12, 273-293.

Askar, A. [1972] Molecular Crystals and Polar Theories of Continua - Experimental Values of Material Coefficients for KNO_3 *Int. J. Engng. Sci.* 10, 293.

Balkanski, M. & Teng, M. K. [1969] *Physics of the Solid State* (eds S. Balakrishna, B. Krishnamurti & B. Ramachaudro). New York: Academic Press.

Coleman, B. D., Markovitz, H. & Noll, W. [1966] *Viscometric Flows of Non-Newtonian Fluids.* New York: Springer.

Coleman, B. D. & Noll, W. [1961] Foundations Of Linear Viscoelasticity *Rev. Modern Phys.* 33, 239-249.

de Gennes, P. G. [1974] *The Physics of Liquid Crystals.* London: Oxford University Press.

Dill, E. H. [1975] Simple Materials with Fading Memory *Continuum Physics* Vol. II (ed A. C. Eringen). New York: Academic Press.

Ericksen, J. L. [1969] Continuum Theory of Liquid Crystals of Nematic Type *Mole. Crys. & Liq. Crys.* 7, 153-164.

Eringen, A. C. [1962] *Nonlinear Theory of Continuous Media.* New York: McGraw Hill

Eringen, A. C. [1964] Simple Micro-fluids *Int. J. Engng. Sci.* 2, 205-217.

Eringen, A. C. [1965] Theory of Micropolar Continua *Developments in Mechanics* (eds T. C. Huang & M. W. Johnson, Jr.). New York: Wiley & Sons.

Eringen, A. C. [1966a] Theory of Micropolar Fluids *J. Math. & Mech.* 16, 1-18.

Eringen, A. C. [1966b] Linear Theory of Micropolar Elasticity *J. Math. & Mech.* 15, 909-924.

Eringen, A. C. [1966c] A Unified Theory of Thermomechanical Materials *Int. J. Engng. Sci.* 41, 179-202.

Eringen, A. C. [1967] *Mechanics of Continua.* New York: Wiley & Sons.

Eringen, A. C. [1967] Linear Theory of Micropolar Viscoelasticity *Int. J. Engng. Sci.* 5, 191-204.

Eringen, A. C. [1968a] Mechanics of Micromorphic Continua *Mech. of Generalized Continua* (ed E. Kroner), 18-35. Berlin: Springer-Verlag.

Eringen, A. C. [1968b] Theory of Micropolar Elasticity *Fracture* Vol. II (ed. H. Liebowitz), 621-729. New York: Academic Press.

Eringen, A. C. [1970a] *Foundations of Micropolar Thermoelasticity*, 1-107. Wien, New York: Springer-Verlag.

Eringen, A. C. [1970b] Balance Laws of Micromorphic Mechanics *Int. J. Engng. Sci.* 8, 819-828.

Eringen, A. C. [1970c] On a Theory of General Relativistic Thermodynamics and Viscous Fluids *A Critical Review of Thermodynamics* (eds E. B. Stuart, B. Gal-Or & A. J. Brainard), 483-503. Baltimore: Mono Book Corp.

Eringen, A. C. [1972a] Theory of Thermomicrofluids *J. Math. Anal. & Appl.* 38, 480-496.

Eringen, A. C. [1972b] Theory of Micromorphic Materials with Memory *Int. J. Engng. Sci.* 10, 623-641.

Eringen, A. C. [1972c] Linear Theory of Nonlocal Elasticity and Dispersion of Plane Waves *Int. J. Engng. Sci.* 10, 425-435.

Eringen, A. C. [1972d] On Nonlocal Fluid Mechanics *Int. J. Engng. Sci.* 10, 561-575.

Eringen, A. C. [1973a] On Rayleigh Surface Waves with Small Wave Length *Letters in Appl. & Engng. Sci.* 1, 11-17.

Eringen, A. C. [1973b] Theory of Nonlocal Electromagnetic Elastic Solids *J. Math. Phys.* 14, 733-740.

Eringen, A. C. [1973c] Linear Theory of Nonlocal Microelasticity and Dispersion of Plane Waves *Letters in Appl. & Engng. Sci.* 1, 129-146.

Eringen, A. C. [1973d] On Nonlocal Microfluid Mechanics *Int. J. Engng. Sci.* 11, 291-306.

Eringen, A. C. [1974] Nonlocal Elasticity and Waves *Continuum Mechanics Aspects on Geodynamics and Rock Fracture Mechanics* (ed P. Thoft-Christensen), 81-105. Dordrecht, Holland: D. Reidel Publishing Co.

Eringen, A. C. [1975] *Continuum Physics, Volume II - Continuum Mechanics of Single-Substance Bodies*. New York: Academic Press.

Eringen, A. C. [1976] Nonlocal Polar Field Theories *Continuum Physics* Vols. III & IV (ed A. C. Eringen). New York: Academic Press.

Eringen, A. C. & Edelen, D. G. B. [1972] On Nonlocal Elasticity *Int. J. Engng. Sci.* 10, 233-248.

Eringen, A. C. & Kafadar, C. B. [1970] Relativistic Theory of Microelectromagnetism *J. Math. Phys.* 11, 1984-1991.

Eringen, A. C. & Kafadar, C. B. [1975] Polar Field Theories *Continuum Physics* Vol. III (ed A. C. Eringen). New York: Academic Press.

Eringen, A. C. & Kim, B. S. [1974] Stress Concentration at the Tip of Crack *Mech. Res. Comm.* 1, 233-237.

Eringen, A. C. & Şuhubi, E. S. [1964] Nonlinear Theory of Simple Microelastic Solids-I *Int. J. Engng. Sci.* 2, 189-203 & II, 4, 389-404.

Filas, R. W., Hajdo, L. E. & Eringen, A. C. [1974] Reorientation of Poly-γ-benzyl Glutamate Liquid Crystals in a Magnetic Field *J. Chem. Phys.* 61, 3037-3038.

Green, A. E. Adkins, J. E. [1960] *Large Elastic Deformations and Nonlinear Continuum Mechanics.* London: Oxford University Press.

Green, A. E. and Zerna, W. [1954] *Theoretical Elasticity.* London: Oxford University Press.

Gray, G. W. [1962] *Molecular Structure and the Properties of Liquid Crystals.* New York: Academic Press.

Grot, R. A. & Eringen, A. C. [1966] Relativistic Continuum Mechanics, I & II *Int. J. Engng. Sci.* 4, 611-638 & 639-370.

Kafadar, C. B. & Eringen, A. C. [1971a] Micropolar Media-I, The Classical Theory *Int. J. Engng. Sci.* 9, 271-305.

Kafadar, C. B. & Eringen, A. C. [1971b] Micropolar Media-II, The Relativistic Theory *Int. J. Engng. Sci.* 9, 307-329.

Kröner, E. [1966] Continuum Mechanics and Range of Atomic Cohesion Forces *Proc. First Nat. Conf. on Fractures*, Vol. I (eds T. Yokobori, T. Kawasaki & J. Swedlow). Sendai, Japan: Japanese Soc. for Strength for Fracture of Materials.

Kunin, A. [1966] Model of Elastic Medium with Simple Structure and Space Dispersion (in Russian) *Prikl. Mat. Mekh.* 30, 542.

Lee, J. C. & Eringen, A. C. [1971a] Wave Propagation in Nematic Liquid Crystals *J. Chem. Phys.* 54, 5027-5034.

Lee, J. C. & Eringen, A. C. [1971b] Alignment of Nematic Liquid Crystals *J. Chem. Phys.* 55, 4504-4508.

Lee, J. D. & Eringen, A. C. [1971c] Boundary Effects of Orientation of Nematic Liquid Crystals *J. Chem. Phys.* 55, 4509-4512.

Lee, J. D. & Eringen, A. C. [1973] Continuum Theory of Smectic Liquid Crystals *J. Chem. Phys.* 58, 4203-4211.

Leslie, F. M. [1969] Continuum Theory of Cholesteric Liquid Crystals *Mole. Crys. & Liq. Crys.* 7, 407-420.

Lockett, F. J. [1972] *Nonlinear Viscoelastic Solids.* New York: Academic Press.

Lodge, A. S. [1964] *Elastic Fluids.* New York: Academic Press.

Maugin, G. A. & Eringen, A. C. [1972a] Deformable Magnetically Saturated Media-I. Field Equations *J. Math. Phys.* 13, 143-155.

Maugin, G. A. & Eringen, A. C. [1972b] Deformable Magnetically Saturated Media-II. Constitutive Theory *J. Math. Phys.* 13, 1334-1347.

Narasimhan, N. L. & Eringen, A. C. [1974] Orientational Effects in Heat-Conducting Nematic Liquid Crystals *Mole. Crys. & Liq. Crys.* 29, 57-87.

Parfitt, R. & Eringen, A. C. [1969] Reflection of Plane Waves from the Flat Boundary of a Micropolar Elastic Half-Space *J. Acous. Soc. Amer.* 45, 1258-1272.

Rivlin, R. S. [1974] The Elements of Non-linear Continuum Mechanics *Continuum Mechanics Aspects of Geodynamics and Rock Fracture Mechanics* (eds P. Thoff-Christensen) 151-175. Dordrecht, Holland: D. Reidel Publishing Co.

Stajanovic, R. [1969] *Mechanics of Polar Continua* (lecture notes). Uduie, Italy: Center of Mechanical Sciences.

Şububi, E. S. [1975] Thermoelastic Solids *Continuum Physics* Vol. II (ed A. C. Eringen). New York: Academic Press.

Truesdell, C. and Noll, W. [1965] *Nonlinear Field Theories of Mechanics* Handbuch der Physick Bd III/3. Berlin: Springer-Verlag.

Truesdell, C. and Toupin, R. [1960] *The Classical Field Theories* Handbuch der Physick Bd III/1. Berlin: Springer-Verlag.

CRITICAL PHENOMENA AND THE RENORMALIZATION GROUP

Franz J. Wegner

Institut fur Theoretische Physik,
Universitat Heidelberg 69 Heidelberg, Germany

ABSTRACT

The recent theory of critical phenomena and the renormalization group as promoted by Wilson is considered on an introductory level. The main emphasis is on the idea of the fixed point Hamiltonian (asymptotic invariance of the critical Hamiltonian under change of the length scale) and the resulting homogeneity laws.

1. CRITICAL BEHAVIOR

A. *Critical points*

The transition from one phase to another like melting or boiling changes the properties of a system discontinuously. (In the beginning we follow widely the introduction of the review article by Fisher [1967]. Compare also the review article by Kadanoff et al. [1967]). Such a phase transition is called a first order transition or discontinuous transition. By varying one or several thermodynamic variables like the temperature, it is frequently possible to follow the coexistence curve so that the two distinct phases become more and more similar until both phases become equal at a certain point. If beyond this point only one homogeneous phase exists and all changes are smooth and continuous, then this point is called a critical point. (Depending on the number of experimentally available variables one may have a critical point, a critical line (λ-line) or even a critical surface, etc. There exist other special points, for example the common end point of a triple point line and three critical lines, called a tricritical point. These points can be described in the framework of this theory, too. They differ from normal critical points by the number of conditions necessary to reach such a point.

Examples of critical points are (a) the termination point of the coexistence curve of a liquid and its vapor (or two phases of different density of a lattice gas like hydrogen in metals) at the critical temperature T_c and pressure p_c, (b) the critical point of separation of mixtures and alloys above (or below) which the components mix without a miscibility gap, (c) the ordering temperature of a homogeneous binary crystal below which one sublattice is primarily occupied by one species.

A second class of systems exhibits domains of magnetic or electric moments of different orientation which vanish at the critical temperature. Examples are (d) ferromagnets with ferromagnetic domains of different orientation whose spontaneous magnetization vanish continuously at the Curie temperature, (e) ferroelectrics with ferroelectric domains whose spontaneous polarization go to zero at the critical temperature, and (f) NH_4 compounds whose electric octupole moments order primarily in one or the other direction below T_c. (g) The alternating spin order of antiferromagnets goes to zero at the Neel point so that two counterphase domains become indistinguishable. Analogously one observes (h) alternating ordering of electric dipole moments in antiferroelectrics and (i) alternating ordering of electric octupole moments in NH_4 compounds.

Thirdly (j) superfluid helium and (k) super-conductors are characterized by a condensate associated with a phase below T_c. This condensate vanishes continuously approaching T_c from below so that domains with different phase cannot be distinguished above T_c. This list does not exhaust the types of critical points observed. But it gives an impression of the variety of phenomena which can be described by the theory of critical phenomena.

To unify the description of critical phenomena one has introduced the concept of the order parameter. For the liquid-vapor transition and other transitions characterized by a difference of densities in both phases (sublattices) the order parameter is the difference between the expectation value of the density of the phase (sublattice) from its value at criticality. For orientational transitions the expectation value of the electric (magnetic) moment (or the difference on the sublattices) serves as the order parameter. In superfluid helium and superconductors the expectation value of the condensate wave function is the order parameter. The amount of the order parameter is (approximately) the same in all phases but it differs in sign, direction and phase, respectively, in different phases (domains).

The field conjugate to the order parameter is often called the symmetry breaking field, since it breaks the symmetry of the Hamiltonian in the case of orientational phase transitions and the transitions to the superfluid and super-conducting state. Without this field the Hamiltonian is invariant under certain rotations of the order parameter or invariant under the change of the phase (gauge transformation). This symmetry breaking field is the magnetic field for ferromagnets, the electric field for ferroelectrics, the chemical potential for the liquid vapor transition, the difference of chemical potentials for mixtures. In several cases the symmetry breaking field is not experimentally accessible as in superfluids and in superconductors. But it is often introduced in theoretical physics for conceptual reasons like the staggered field in antiferromagnets and antiferroelectrics.

This unified description allows us to restrict to one class of systems in explaining the main features of critical phenomena, provided we neglect a number of peculiar features of certain systems. Two features we will often neglect are (i) the quantum mechanic (or discrete) nature of the microscopic origin of many phase transitions (superfluid He, superconductors, spin and exchange interaction in magnets, etc.) Since critical phenomena become apparent on a macroscopic scale, it is assumed that the commutators can be neglected and the order parameter can be handled like a continuous classical variable. (ii) In many cases we will neglect long-range interactions or the long-range part of these interactions. Therefore we will neglect dipolar interactions and the coupling of the interaction to lattice distortions which induce long range interactions.

We will mainly use the magnetic language. Thus we will discuss the critical behavior of a ferromagnet consisting of classical spins on a rigid lattice for which the exchange interaction dominates so that the dipolar interaction can be neglected

B. *Critical exponents - the homogeneity assumption*

The first theory to explain the critical behavior of ferromagnets was the molecular field theory by P. Weiss [1907].(The first theory which described a phase transition (for a liquid-vapor system) is due to van der Waals [1873]).

According to this theory the spontaneous magnetization m is zero above T_c and goes to zero from below like $\sqrt{T_c - T}$. The susceptibility diverges like $|T - T_c|^{-1}$ from below and above T_c (see Fig. 1) and the specific heat show a finite

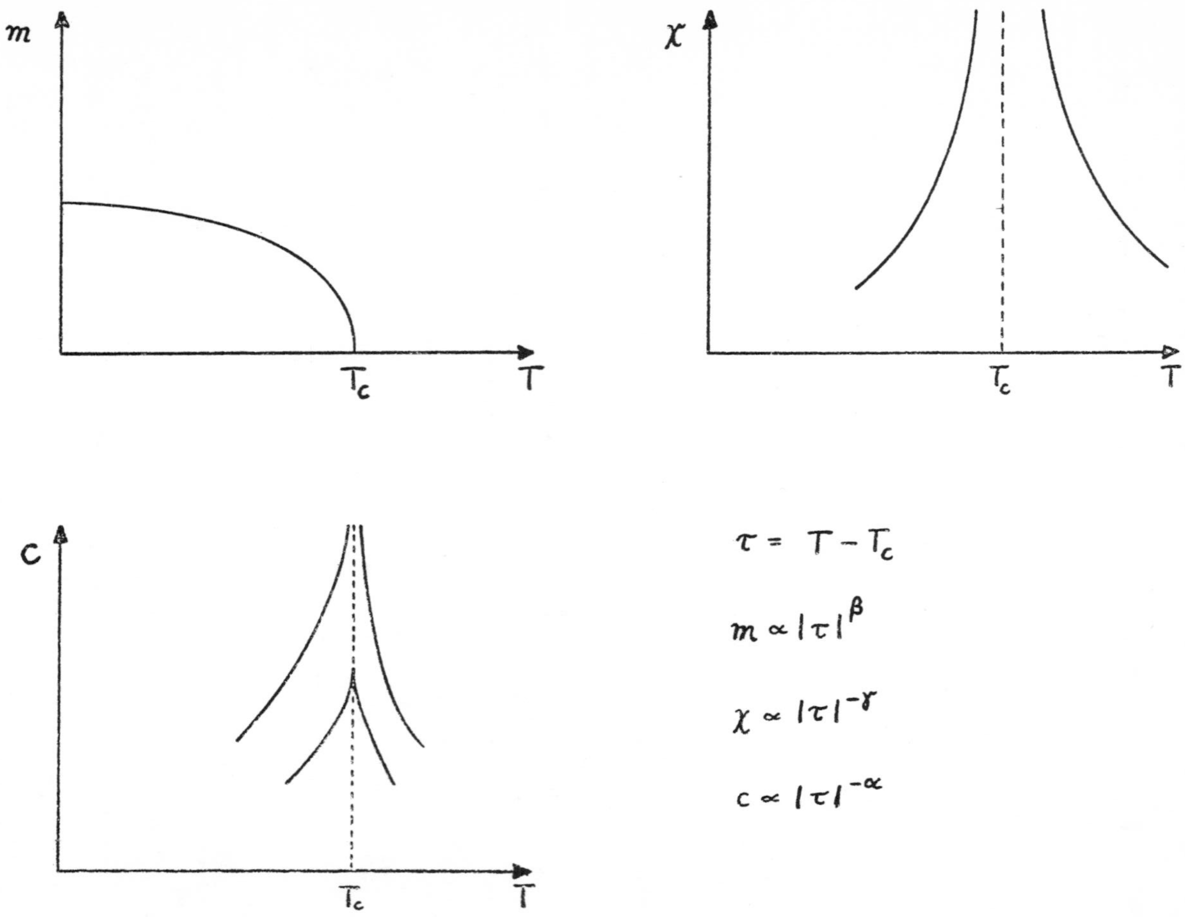

$$\tau = T - T_c$$

$$m \propto |\tau|^{\beta}$$

$$\chi \propto |\tau|^{-\gamma}$$

$$c \propto |\tau|^{-\alpha}$$

Fig. 1. The schematic behavior of the spontaneous magnetization, susceptibility and specific heat near T_c.

jump at T_c. Experimentally however one finds $m \propto (T_c - T)^{\beta}$ with $\beta \approx 1/3$ for the spontaneous magnetization, $\chi \propto |T_c - T|^{-\gamma}$ with $\gamma \approx 4/3$ and a singular contribution to the specific heat like $c_{sing} \propto |T_c - T|^{-\alpha}$ with α close to zero. For negative α the specific heat shows a cusp, for positive α it diverges. The exponents α, β and γ are called critical exponents. The deviation of the molecular field exponents from the experimental critical exponents has led to the search of soluble models. Unfortunately most models (approximations) lead back to the molecular field behavior. Two models which give different sets of critical exponents are the spherical model and the two-dimensional Ising model. The exponents are listed in Table 1. None of these models give exponents which are close to the experimentally observed exponents. The reason is that the molecular field theory completely neglects the critical fluctuations (apart from the homogeneous component) which leads to a γ which is too small; the spherical model overestimates the critical fluctuations which leads to a γ which is too large. The two-dimensional Ising model describes the fluctuations properly. However the dimensionality of the system plays an important role in critical phenomena so that the two-dimensional Ising model does not yield a reasonable approximation for the three-dimensional Ising model.

Apart from some other two-dimensional models (F-model, KDP-model, eight-vertex-model), there are no exactly soluble models available. Therefore one has tried a different approach to determine critical exponents by means of series expansions. One expands for example the susceptibility or the specific heat of a model like the Ising model in powers of the inverse temperature $\beta = (k_B T)^{-1}$, *assumes* that the quantity considered shows a power law behavior close to T_c and analyzes the series accordingly. (We hope that the reader does not get confused since as usual β is used for the critical exponent of the order parameter and the inverse temperature.) This yields estimates for the critical exponents listed in the last columns of Table 1 for three models: The Ising model (a model of spins S with two states $S = \pm 1$), the XY-model (a model of planar spins S, that is, spins with two components $S_x = \cos\phi$, $S_y = \sin\phi$) and the classical Heisenberg model (a model of three-dimensional [classical] vectors S with $S^2 = 1$). The spins are located at the sites of a lattice and interact via an (isotropic) short-range (in most cases nearest neighbor) interactions. Low temperature expansions are only available for the Ising model. Therefore β is quoted only for the Ising model. One can estimate the low-temperature exponents for the specific heat and the susceptibility of the Ising model. They are slightly different from the high-temperature exponents. Since it is hard to estimate the accuracy of the exponents determined, it is hard to decide whether high and low-temperature exponents are equal within the error bars. One finds that the exponents determined from the expansions are quite close to the experimentally observed

Power law	Exponent	Molecular Field Approxim.	Spherical Model d=3	Ising Model d=2	Experiments d=3	Hightemperature expansions d=3 n=1	n=2	n=3
$m \propto \|\tau\|^{\beta}$	β	1/2	1/2	1/8	≈1/3	.31		
$\chi \propto \|\tau\|^{-\gamma}$	γ	1	2	7/4	≈4/3	1.25	1.32	1.38
$c_{sing} \propto \|\tau\|^{-\alpha}$	α	0 (disc.)	-1 (kink)	0 (log)	≈0	.13	.00	-.10

Table 1. Critical exponents of various models.

ones. But unfortunately one does not learn from these expansions why the systems exhibit these broken power laws near T_c. It is the aim of this paper to review at an introductory level the ideas which provide an understanding of the critical behavior.

A first step to link different aspects was the homogeneity assumption by Widom [1965]. We bring a modified version of it. (Widom's assumption included the possibility of logarithmic singularities which will not be considered in this section.) Widom assumes that the free energy can be separated as a function of the magnetic field h and the temperature difference $\tau = T - T_c$ into a regular and a singular part

$$F = F_{reg}(\tau) + F_{sing}(\tau,h), \tag{1.1}$$

where the singular part is responsible for the critical behavior. He assumes that the singular part is a homogeneous function of the variables τ and h, that is

$$F_{sing}(\tau,h) = |\tau|^{2-\alpha} f_{\pm}\left(\frac{h}{|\tau|^{\Delta}}\right) \tag{1.2}$$

where the $\pm$ denotes that the function is different for positive and negative τ. Δ is called the gap exponent. Homogeneity means that multiplying τ by a factor c, and h by a factor c^{Δ} multiplies the function by a factor $c^{2-\alpha}$.

Let us discuss some consequences. We obtain the specific heat by differentiating F twice with respect to α. (A number of factors $k_B T$ is missing which however does not affect the critical exponents.)

This gives the singular part of the specific heat at constant vanishing field h

$$c_{sing} \propto |\tau|^{-\alpha} f_{\pm}(0) \tag{1.3}$$

which was the reason for calling the exponent in Eq.(1.2) $2-\alpha$. The magnetization is obtained from eq.(1.2) by differentiating with respect to h

$$m = -|\tau|^{2-\alpha-\Delta} f'_{\pm}\left(\frac{h}{|\tau|^{\Delta}}\right) \tag{1.4}$$

At $h = 0$ this leads to

$$m = -|\tau|^{\beta} f'_{\pm}(0) \tag{1.5}$$

with

$$\beta = 2 - \alpha - \Delta. \tag{1.6}$$

Differentiating twice with respect to h we obtain

$$\chi = -|\tau|^{2-\alpha-2\Delta} f''_{\pm}\left(\frac{h}{|\tau|^{\Delta}}\right) \tag{1.7}$$

which yields

$$\gamma = \alpha + 2\Delta - 2. \tag{1.8}$$

From eqs.(1.6) and (1.8) we find a relation between the exponents α, β, γ

$$\alpha + 2\beta + \gamma = 2. \tag{1.9}$$

A look at Table 1 shows that this relation is fulfilled for all listed sets of exponents. We note that from eqs.(1.6) and (1.8) we obtain

$$\Delta = \beta + \gamma. \tag{1.10}$$

Then eq.(1.4) can be easily cast in the form

$$\frac{m}{|\tau|^{\beta}} = - f'_{\pm}\left(\frac{h}{|\tau|^{\Delta}}\right) \tag{1.11}$$

Solving with respect to $h/|\tau|^{\Delta}$ yields

$$\frac{h}{|\tau|^{\Delta}} = g_{\pm}\left(\frac{m}{|\tau|^{\beta}}\right) \tag{1.12}$$

with some function g, which can be written

$$\frac{h}{m|\tau|^{\gamma}} = w_{\pm}\left(\frac{m}{|\tau|^{\beta}}\right) \tag{1.13}$$

with $w(x) = g(x)/x$. Thus $h/|\tau|^{\Delta}$ should be a function of $m/|\tau|^{\beta}$ only. In Fig.2 data (cf. Ho and Litster [1969]) of the magnetization m of $CrBr_3$ as a function of the two variables, temperature and field, are plotted in the variables $m/|\tau|^{\beta}$ and $h/(m/|\tau|^{\gamma})$. If the homogeneity assumption would not hold, the data points would be scattered in the whole plot. Since the data follow the homogeneity assumption, they lie on two lines corresponding to the behavior above and below T_c.

In the following section we will show how the homogeneity relation can be derived.

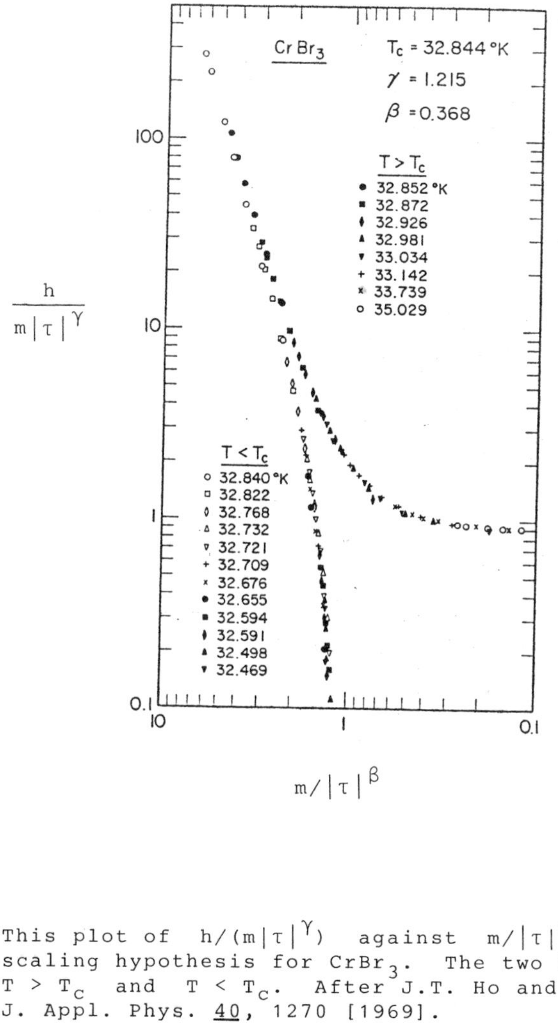

Fig.2. This plot of $h/(m|\tau|^{\gamma})$ against $m/|\tau|^{\beta}$ confirms the scaling hypothesis for $CrBr_3$. The two branches are for $T > T_c$ and $T < T_c$. After J.T. Ho and J.D. Litster, J. Appl. Phys. 40, 1270 [1969].

2. RENORMALIZATION GROUP EQUATION AND SCALING

A. *Motivation*

A hint on how the critical state can be characterized can be obtained from the correlation functions. Let us consider the auto correlation function of the spins $S_o(r)$. From what one knows from exactly solvable systems this correlation function decays at criticality with a power law for large distances

$$<S_o(0)S_o(r)>_{crit} = \frac{c}{r^{d-2+\eta}} \tag{2.1}$$

where η is a new critical exponent and d the dimensionality of the system. η describes the deviation from the Ornstein-Zernicke-behavior of the correlation function. Let us now consider the same ferromagnet under a different length scale. To accomplish this we divide the sample into cubic cells of length b lattice spacings in each direction. Then the magnetization of a cell

$$s = \sum_{cell} S_o(r'), \tag{2.2}$$

obeys asymptotically

$$<s(0)s(r)> = \frac{c\mathbf{b}^{2d}}{r^{d-2+\eta}} \tag{2.3}$$

since each cell contains b^d spins. Now we change the length scale by a factor b and the scale for the magnetization by a factor $b^{(d+2-\eta)/2}$

$$r = \mathbf{b}R, \qquad s(r) = \mathbf{b}^{(d+2-\eta)/2} s_1(R) \tag{2.4}$$

Then we obtain the asymptotic behavior of our new spin variables

$$\langle S_1(0)\ S_1(R)\rangle = \frac{c}{R^{d-2+\eta}} \tag{2.5}$$

Therefore the correlation function is invariant under the change of the scale (2.4). This invariance of the correlation function suggests that the effective interaction at criticality is invariant with respect to the change of the length scale.
We call the procedure which changes the scale of the Hamiltonian (effective interaction) renormalization group (RG) procedure and the corresponding transformation is called RG transformation.
In the remainder of this section we outline some requirements and properties of RG transformation and derive the scaling of the free energy.

B. *Properties of RG transformation*

We denote the Hamiltonian function $\mathcal{H}$ and the free energy $\mathcal{F}$.
We introduce

$$H = \beta\mathcal{H} \qquad F = \frac{\beta}{V}\mathcal{F}, \qquad \beta = \frac{1}{k_B T} \tag{2.6}$$

where V is the volume of the system. For simplicity's sake we call H and F Hamiltonian and free energy, resp.

$$-F = \frac{1}{V}\,\ell n\ \mathrm{tr}\ \exp(-H). \tag{2.7}$$

The RG transformation consists of

(i) a change of the length scale by a factor $b = e^{\ell}$ in all linear dimensions (we leave the partition function $Z = \text{trace}\exp(-H)$ invariant). Since the volume shrinks by a factor $e^{-d\ell}$ we obtain

$$F_o = e^{-d\ell} F_\ell \tag{2.8}$$

(ii) a transformation and/or elimination of the spin variables S which leaves the free energy invariant. The transformation shall not generate long-range interactions. The new Hamiltonian H_ℓ has to be comparable with the original Hamiltonian H_o (same Hilbert or function space). This demands an extension of the system of the original volume for finite systems. The RG transformation transforms H_o into H_ℓ

$$H_\ell = R_\ell(H_o) \tag{2.9}$$

$$F(H_o) = e^{-d\ell} F(H_\ell) \tag{2.10}$$

C. *RG Equations*

Here we list a number of ways to construct RG equations which transform Hamiltonians:

(i) Wilson's recurrence relation (approximation) (cf. Wilson [1971]). Numerical solutions for $d = 3$ (cf. Grover, Kadanoff, Wegner [1972], Grover [1972], Swift and Grover [1974]). Expansion in $\varepsilon = 4-d$ (cf. Wilson and

Fisher [1972], Fisher and Pfeuty [1972], Wegner [1972], Wilson [1972], Brezin, Guillou, Zinn-Justin [1973], Houghton and Wegner [1974]).

(ii) Wilson's differential RG equation with smooth momentum cut-off (cf. Wilson and Kogut [1974]), generalization and applications (cf. Wegner [1974], Shukla and Green [1974], [1975], Golner and Riedel [1975], Rudnick [1975], Wegner [to appear]).

(iii) Differential RG equation with sharp momentum cut-off (generates long-range interactions): Wegner and Houghton [1973]

(iv) Aharony's method (cf. Aharony and Fisher [1973], Aharony [1973], Bruce, Droz and Aharony [1974]).

(v) Two-(three)-dimensional Ising models: Niemeijer and van Leeuwen [1973],[1974], Nauenberg and Nienhuis [1974], Kadanoff and Houghton [1975]

There are other varieties of the RG which transform the correlation functions (Callan-Symanzik-equation, compare the review by Zinn-Justin [1973]). To give an idea of the construction of a RG let us consider that by Niemeijer and van Leeuwen ([1973], [1974]) for the Ising model on a triangular lattice. They divide the lattice into triangles i each of which contains three spins S_{i1}, S_{i2}, S_{i3}. They define the new Ising variables

$$S_i' = \mathrm{sign}(S_{i1} + S_{i2} + S_{i3}) \tag{2.11}$$

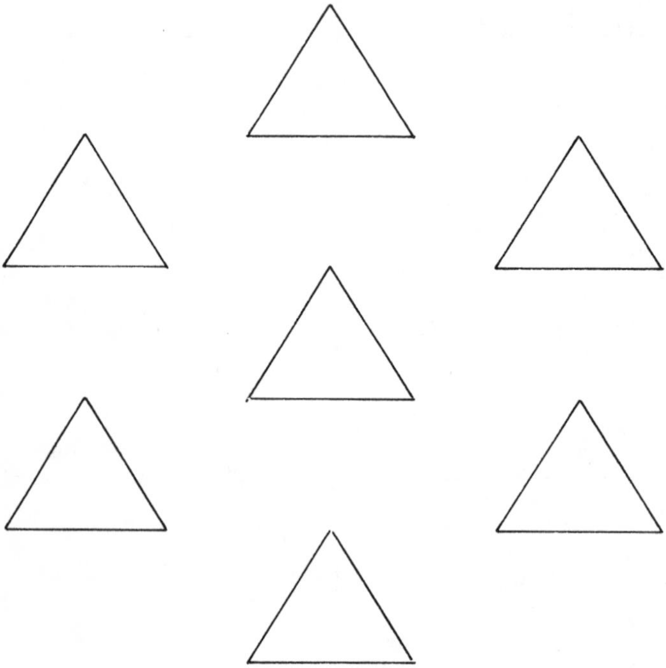

Fig.3. The triangular lattice used by Niemeijer and van Leeuwen [1973], [1974].

and calculate (approximately) a new Hamiltonian $H'\{S'\}$ and the original Ising Hamiltonian $H\{S\}$ according to

$$\exp(-H'\{S'\}) = \Sigma \exp(-H_o\{S\}) \tag{2.12}$$

where the sum runs over all configurations $\{S\}$ which fulfill equation (2.11). The spins S' form a triangular lattice again (Fig. 3).

D. *Fixed point, classification of operators*

In the Wilson theory of critical phenomena the following two assumptions are made:

(i) It is assumed that a fixed point Hamiltonian H^* exists

$$R_\ell(H^*) = H^* \tag{2.13}$$

This is a Hamiltonian which maps into itself.

(ii) It is assumed that for a critical Hamiltonian the limit

$$\lim_{\ell\to\infty} H_\ell = H^* \tag{2.14}$$

approaches a fixed point H^*.

We assume in the following that the eigenoperators form a complete set of operators so that any Hamiltonian H_o can be expanded

$$H_o = H^* + \Sigma \mu_i O_i. \tag{2.15}$$

Then we obtain in linear order in μ

$$H_\ell = H^* + \Sigma\, \mu_i e^{y_i \ell} O_i \tag{2.16}$$

Corresponding to the eigenvalues y one distinguishes

$$\begin{array}{ll} y > 0 & \textit{relevant} \text{ operator} \\ y = 0 & \textit{marginal} \text{ operator} \\ y < 0 & \textit{irrelevant} \text{ operator} \end{array} \tag{2.17}$$

From equation (2.16) we find immediately that at the critical point the fields (in high energy physics sources) μ_i of all relevant operators have to vanish.

Depending on the nonlinear terms marginal operators may act as relevant, irrelevant, and substantially marginal operators (as in the eight-vertex-model), resp.

Since the volume V shrinks by a factor $e^{-d\ell}$ the factor μ_o in the constant $\mu_o V$ grows like $e^{d\ell}$. Therefore there is a *special* operator, the volume V, with $y_o = d$. However the addition of a constant to the Hamiltonian does not change its critical behavior. Therefore $\mu_o = 0$ is not necessary for criticality. This is the origin of the regular part of the free energy.

The type of the critical behavior depends on the number of symmetry conserving relevant operators. (Symmetry conserving means that the symmetry of the Hamiltonian is conserved, it does not exclude a spontaneously broken symmetry of the system). Let us expand

$$\mathcal{H} = \Sigma\, \mu_i^o O_i \tag{2.18}$$

$$H^* = \Sigma\, \mu_i^* O_i \tag{2.19}$$

then we obtain

$$\mu_i = \beta\mu_i^o - \mu_i^* . \tag{2.20}$$

For a normal critical point one has one relevant symmetry conserving operator (apart from V) O_E which determines the critical temperature

$$\mu_E \equiv \tau = \beta\mu_E^o - \mu_E^* \tag{2.21}$$

Crudely speaking O_E is proportional to the Hamiltonian minus its expectation value at the critical point. At a critical point one has two relevant symmetry conserving operators (apart from O_o) and consequently two conditions for criticality.

E. *Scaling of the free energy*

Within a simplified picture (Kadanoff's cell model [1966]) we consider only two operators O_E and the magnetization O_h

$$H_o = H^* + \tau O_E + hO_H \tag{2.22}$$

which yields

$$H_\ell = H^* + \tau e^{y_E \ell} O_E + he^{y_h \ell} O_h \tag{2.23}$$

$$F(\tau,h) = e^{-d\ell} F(\tau e^{y_E \ell},\ he^{y_h \ell}) \tag{2.24}$$

we choose τ by

$$|\tau|\, e^{y_E \ell} = 1 \tag{2.25}$$

and obtain Widom's scaling law (1.2)

$$F(\tau,h) = |\tau|^{d/y_E} F(\pm 1, \frac{h}{|\tau|^{y_h/y_E}}) \tag{2.26}$$

with

$$d/y_E = 2 - \alpha \qquad\qquad y_h/y_E = \Delta . \tag{2.27}$$

Normally one has an infinite number of perturbations O_i in equation (2.22). To study their effect on the scaling law we add at least one further operator pars pro toto

$$H_o = H^* + \tau O_E + hO_h + \mu_i O_i \tag{2.28}$$

and obtain

$$H_\ell = H^* + \tau e^{y_E \ell} O_E + h e^{y_h \ell} O_h + \mu_i e^{y_i \ell} O_i \tag{2.29}$$

$$F(\tau,h,\mu_i) = |\tau|^{d/y_E} F(\pm 1, \frac{h}{|\tau|^{y_h/y_E}}, \frac{\mu_i}{|\tau|^{y_i/y_E}}). \tag{2.30}$$

We are interested in the critical behavior, that is in the limit $\tau \to 0$

$$\lim_{\tau\to 0} \frac{\mu_i}{|\tau|^{y_i/y_E}} \to \begin{cases} 0 & \text{for } y_i < 0 \quad \text{or} \quad \mu_i = 0 \\ \pm\infty & \text{for } y_i > 0 \quad \text{and} \quad \mu_i \neq 0 \end{cases} \tag{2.31}$$

If O_i is relevant ($y_i > 0$) then μ_i has explicitly to be taken into account. For irrelevant operators the term $\mu_i/|\tau|^{y_i/y_E}$ can be neglected if F can be expanded in powers of μ_i. Note that the right hand side of eq.(2.31) contains the free energy well apart from the critical point. The irrelevant operator yields a correction to scaling

$$F = |\tau|^{d/y_E} F(\pm 1, \frac{h}{|\tau|^{y_h/y_E}}, 0) + |\tau|^{\frac{d-y_i}{y_E}} \mu_i F'(\pm 1, \frac{h}{|\tau|^{y_h/y_E}}, 0) + \ldots \tag{2.32}$$

as observed in superfluid He (Ahlers [1973]). If F cannot be expanded in powers of μ_i, then Fisher's idea of the anomalous dimension of the vacuum might apply (cf. Fisher [1973]).

F. *Correlation length*

The correlation functions decay apart from criticality like $e^{-r/\xi}$. The correlation length ξ changes under the RG transformation due to the transformation of the length scale

$$\xi(\tau e^{y_E \ell}, h e^{y_h \ell}) = e^{-\ell} \xi(\tau, h) \tag{2.33}$$

which implies

$$\xi(\tau, h) = |\tau|^{-\nu} (\pm 1, \frac{h}{|\tau|^{\Delta}}) \tag{2.34}$$

with the critical exponent

$$\nu = 1/y_E \tag{2.35}$$

3. CALCULATION OF CRITICAL EXPONENTS

A. *An approximate RG equation*

We shortly derive Wilson's approximate RG equation. The interaction of N spins in a volume V may be approximately described by a Landau type of interaction

$$H_o = \int d^d x \, ((\nabla M)^2 + r_o M^2(x) + u_o M^4(x) - hM(x)) \tag{3.1}$$

where r_o depends crucially on the temperature. It is positive well above T_c and negative well below T_c. Let us separate $M(x)$ in a contribution $M'(x)$ varying slowly in space and a fluctuating part (short wave lengths)

$$M(x) = M'(x) + \Sigma \, m_i M_i(x) \tag{3.2}$$

The N degrees of freedoms are divided into $Ne^{-d\ell}$ determining $M'(x)$ and $N(1-e^{-d\ell})$ variables m_i. The functions $M_i(x)$ are fixed. We expand H_o in the variables m_i

$$\begin{aligned} H_o = & \int d^d x ((\nabla M')^2 + r_o M'^2 + u_o M'^4 - hM') \\ & + \Sigma m_i \int d^d x \, [2(\nabla M')(\nabla M_i(x)) + 2r_o M'(x) M_i(x) \\ & + 4u_o M'^3(x) M_i(x) - hM_i(x)] \\ & + \Sigma m_i m_j \int d^d x [\nabla M_i(x) \nabla M_j(x) + r_o M_i(x) M_j(x) \\ & + 6u_o M'^2(x) M_i(x) M_j(x)] + \ldots \end{aligned} \tag{3.3}$$

Within a rough approximation we require:

(i) The functions $M_i(x)$ are well localized around x_i, so that $M'(x)$ does not change appeciably within the region $M_i(x) \neq 0$,

(ii) $\int M_i(x) d^d x = 0$

(iii) The functions $M_i(x)$ do not overlap

$$\int M_i(x) M_j(x) d^d x = \delta_{ij}$$
$$\int \nabla M_i(x) \nabla M_j(x) d^d x = c\delta_{ij} \tag{3.4}$$

Then equation (3.3) reads

$$H_o = \int d^d x((\nabla M')^2 + r_o M'^2 + u_o M'^4 - hM')$$
$$+ \sum_i m_i^2 (c + r_o + 6u_o M'^2(x_i)). \tag{3.5}$$

Now we may perform the elimination of the variables m_i

$$\exp(-H') = \int \pi dm_i \exp(-H_o) \tag{3.6}$$

which yields

$$H' = \int d^d x((\nabla M')^2 + r_o M'^2 + u_o M'^4 - hM')$$
$$+ \frac{1}{2} \sum_i \ell n(c + r_o + 6u_o M'^2(x_i)) \tag{3.7}$$

Since we eliminate $N(1-e^{-d\ell}) \approx Nd\ell$ terms we may replace the sum

$$\sum_i \approx \frac{d\ell}{\upsilon} \int d^d x \tag{3.8}$$

where υ is the volume per spin

$$H' = \int d^dx[(\nabla M')^2+r_oM'^2+u_oM'^4-hM'+\frac{\ell d}{2\upsilon}\ln(c+r_o+6u_oM'^2(x))] \tag{3.9}$$

which for small u_o can be expanded

$$H' = \int d^dx[(\nabla M')^2-hM'+(r_o+\frac{3u_o\ell d}{\upsilon(c+r_o)})M'^2+(u_o-\frac{9u_o^2\ell d}{\upsilon(c+r_o)^2})M'^4] \tag{3.10}$$

The numbers of variables per volume in H' is less than in H_o. Therefore we have to perform a scale transformation

$$x = e^{\ell}x_1 . \tag{3.11}$$

To reproduce the factor 1 in front of the $(\nabla M)^2$ term we perform a scale transformation for the magnetization

$$M' = e^{\frac{1}{2}(2-d)\ell}M_1 \tag{3.12}$$

which yields for small ℓ

$$\begin{aligned} H_\ell = \int d^dx_1((\nabla M_1)^2&+(r_o+\frac{3u_o\ell d}{\upsilon(c+r_o)}+2\ell r_o)M_1^2-h(1+(\frac{d}{2}+1)\ell)M_1 \\ &+(u_o-\frac{9u_o^2\ell d}{\upsilon(c+r_o)^2}+(4-d)\ell u_o)M_1^4+\ldots) \end{aligned} \tag{3.13}$$

We obtain differential RG equations for the parameters h_1r and u:

$$\frac{dh}{d\ell} = (\frac{d}{2}+1)h \tag{2.14a}$$

$$\frac{dr}{d\ell} = 2r+\frac{3ud}{\upsilon(c+r)} = 2r+\frac{d}{\upsilon c^2}(3cu-3ur+\ldots) \tag{3.14b}$$

$$\frac{du}{d\ell} = (4-d)u - \frac{9u^2 d}{\upsilon(c+r)^2} = (4-d)u - \frac{d}{\upsilon c^2}(9u^2+\ldots) \qquad (3.14c)$$

B. *Solution of the RG equations*

We calculate the fixed points from equations (3.14). We obtain

$$h^* = 0. \qquad (3.15)$$

In equations (3.14b) and (3.14c) we keep all terms to order ur and u^2

$$2r^* + \frac{d}{\upsilon c^2}(3cu^* - 3u^* r^*) = 0 \qquad (3.16)$$

$$(4-d)u^* - \frac{d}{\upsilon c^2} 9u^{*2} = 0 \qquad (3.17)$$

The second equation has two solutions

the trivial fixed point (tfp) solution $u^* = 0$ (3.18)

the nontrivial fixed point (ntfp) solution $u^* = \frac{\upsilon c^2(4-d)}{9d}$ (3.19)

It has been assumed in the beginning that u^* should be a small number. This is obviously the case for small $\varepsilon = 4-d$

$$(\text{ntfp}) \qquad u^* = \frac{\upsilon c^2 \varepsilon}{36} \qquad (3.20)$$

From equation (3.16) we find

$$(\text{tfp}) \qquad r^* = 0 \qquad (3.21)$$

$$(\text{ntfp}) \qquad r^* = -\frac{c\varepsilon}{6} \qquad (3.22)$$

After determination of the fixed point we may linearize the equations (3.14) to obtain the eigenperturbations

$$\frac{dh}{d\ell} = (3 - \frac{\varepsilon}{2})h, \qquad (3.23)$$

$$\frac{d\Delta r}{d\ell} = (2 - \frac{12u^*}{\upsilon c^2})\Delta r + \frac{4}{\upsilon c^2}(3c - 3r^*)\Delta u, \qquad (3.24)$$

$$\frac{d\Delta u}{d\ell} = (\varepsilon - \frac{72u^*}{\upsilon c^2})\Delta u. \qquad (3.25)$$

Easily one obtains the eigenvalues

$$y_h = 3 - \frac{\varepsilon}{2} \qquad (3.26)$$

$$y_E = 2 - \frac{12u^*}{\upsilon c^2} \qquad (3.27)$$

$$y_u = \varepsilon - \frac{72u^*}{\upsilon c^2} \qquad (3.28)$$

With equations (3.18) and (3.20) we obtain

$$\text{(tfp)} \quad y_E = 2 \qquad y_u = \varepsilon \qquad (3.29)$$

$$\text{(ntfp)} \quad y_E = 2 - \frac{\varepsilon}{3} \qquad y_u = -\varepsilon \qquad (3.30)$$

In $d = 4-\varepsilon$ dimensions, $\varepsilon > 0$, the trivial fixed point has two relevant even operators, the nontrivial fixed point has one relevant even operator. Thus we expect that the normal critical behavior is described by the nontrivial fixed point.

One obtains the exponents

$$\alpha = \frac{1}{6}\varepsilon, \qquad \beta = \frac{1}{2} - \frac{1}{6}\varepsilon, \qquad 2\nu = \gamma = 1 + \frac{1}{6}\varepsilon. \tag{3.31}$$

These critical exponents go into the direction expected from the exponents obtained from high temperature and low temperature expansions.

C. *Isotropic n-vector model*

Extensive calculations have been performed for the isotropic n-vector model. This is an isotropic model for n-dimensional vectors $\vec{m}$. $n = 1,2,3$, and ∞ corresponds to the Ising model, XY-model, isotropic Heisenberg-model, and the spherical model (Stanley [1968]) respectively. One distinguishes three types of calculations:

(i) Approximate calculations. Wilson's recurrence relation (cf. Wilson [1971]) can be used to calculate numerically the critical exponents. Golner and Riedel [1975] truncated exact RG equations to calculate critical exponents. They are shown in Table 2.

(ii) The critical exponents can be expanded around dimensionality as before. Unfortunately however the series seem to be asymptotically divergent. As a thumb-rule one finds that the exponents in order ε^2 yield a good approximation.

$$\alpha = \frac{(4-n)}{2(n+8)}\varepsilon + \ldots, \qquad \beta = \frac{1}{2} - \frac{3}{2(n+8)}\varepsilon + \ldots, \qquad \gamma = 1 + \frac{(n+2)}{2(n+8)}\varepsilon + \ldots \tag{3.32}$$

Table 2. Critical exponents as obtained by various methods for d=3

		n=1	n=2	n=3	Ref.
α	high temp. exp.	.13	.00	-.10	Jasnow and Wortis [1968] compare Fisher [1967], Kadanoff et al. [1967].
	r. r. numerical	.17	.07	-.04	Wilson [1971] Grover, Kadanoff, Wegner [1972], Grover [1972], Swift and Grover [1974].
	Trunc. RG eq.	.10			Golner and Riedel [1975].
	in $O(\varepsilon)$	.17	.10	.05	Wilson & Fisher [1972] Fisher & Pfeuty [1972] Wegner [1972] Wilson [1972] Brezin, Guillou & Zinn-Justin [1973], Houghton & Wegner [1974].
	in $O(\varepsilon^2)$	.08	-.02	-.10	" " "
	in $O(\varepsilon^3)$	.20	+.08	.01	" " "
	experiment	.16	-.02	-.14	Ahlers [1973], [1974]
γ	high temp. exp.	1.25	1.32	1.38	Same as above
	r. r. numerical	1.22	1.29	1.36	Same as above
	trunc. RG eq.	1.25			Same as above
	in $O(\varepsilon)$	1.17	1.20	1.23	Same as above
	in $O(\varepsilon^2)$	1.24	1.30	1.35	Same as above
	in $O(\varepsilon^3)$	1.19	1.26	1.32	Same as above

(iii) The critical exponents can be expanded in powers of $1/n$. For $n = \infty$ one obtains the critical exponents for the spherical model (cf. Stanley [1968])

$$\alpha = \frac{d-4}{d-2}, \qquad \beta = \frac{1}{2}, \qquad \gamma = \frac{2}{d-2} \tag{3.33}$$

One can perform a systematic expansion (cf. Ferrell and Scalapino [1972], Abe [1973], Abe and Hikami [1973], Hikami [1973], Ma [1972],[[1973], Fisher, Ma and Nickel [1972], Suzuki [1973]) around this limit which yields for $d = 3$

$$\alpha = -1 + \frac{32}{\pi^2 n} + \ldots, \qquad \gamma = 2 - \frac{24}{\pi^2 n} + \ldots \tag{3.34}$$

These numbers are not yet good approximations although they tend into the correct direction. One has to wait for terms in order $1/n^2$. Thus we have expansions around $n = \infty$ and $d = 4$. Moreover, for $n = -2$ one has the exact result (Balian and Toulouse [1973], Fisher [1973])

$$\gamma = 1, \qquad \alpha = \frac{1}{2}(4-d). \tag{3.35}$$

Further one knows that for $d = 2$, $n > 1$ there is no spontaneous magnetization (cf. Mermin and Wagner [1966]). In Figs. 4 and 5 we plot α and γ as functions of d and n.

Besides the isotropic n-vector-model there are other types of critical behavior in magnetic systems:

(i) cubic-short range

(ii) isotropic dipolar

(iii) uniaxial dipolar

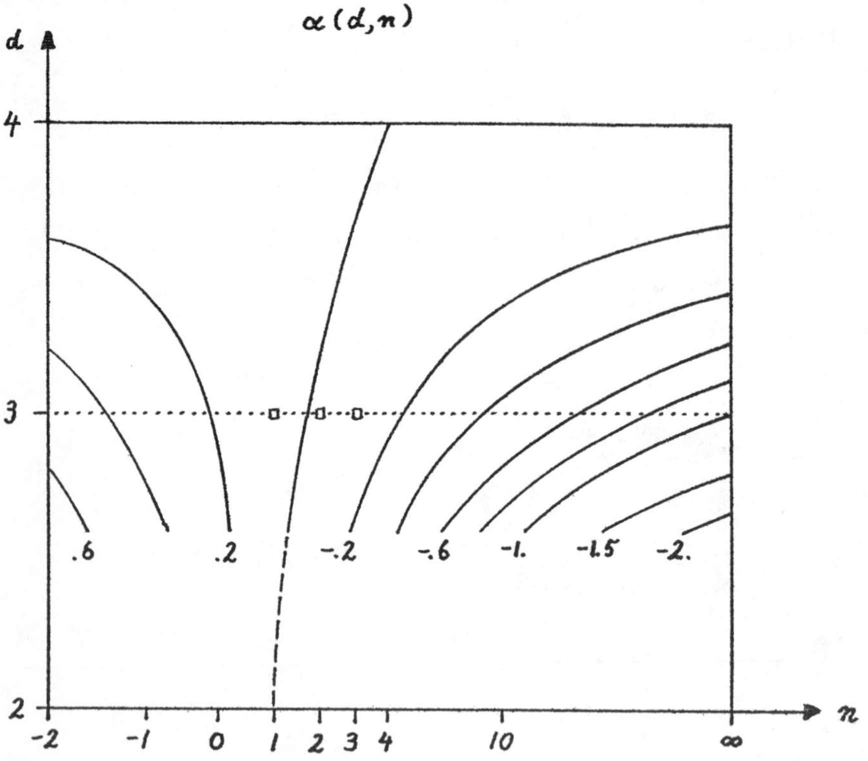

Fig.4. The exponent α(d,n)

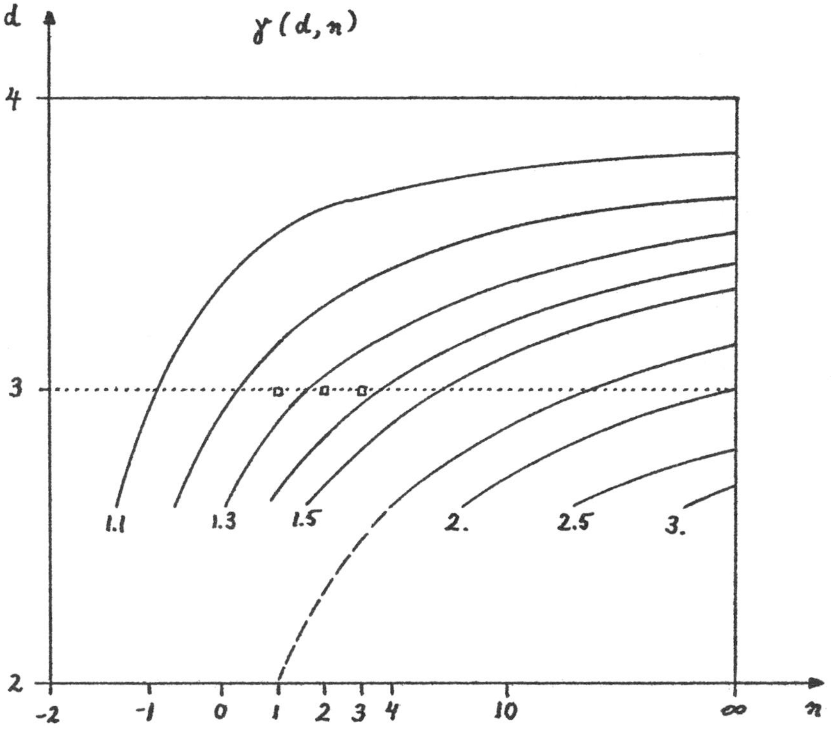

Fig.5. The exponent γ(d,n)

(iv) tricritical behavior

(v) short-range magnets on elastic lattices

For a survey and references see the review article by Fisher [1974]. It is noteworthy that the isotropic n-vector-model and the models (i), (ii), and (v) have nontrivial fixed points branching off at dimensionality 4, whereas the systems (iii) and (iv) show molecular field behavior with logarithmic corrections since the nontrivial fixed point branches off at $d = 3$.

REFERENCES

Abe, R. [1972] Prog. Theor. Phys. 48, 1414; [1973] 49, 113.

Abe. R. and Hikami, S. [1973] Prog. Theor. Phys. 49, 442.

Aharony, A. [1973] Phys. Rev. B8, 3342, 3349, 3358, 3363.

Aharony, A. and Fisher, M.E. [1973] Phys. Rev. B8, 3323.

Ahlers, G. [1973] Phys. Rev. A8, 530

Ahlers, G. Kornblit, A. & Salamon, M.B. [1974] Phys. Rev. B9, 3932.

Balian, R. and Toulouse, G. [1973] Phys. Rev. Lett. 30, 544.

Brezin, E., Le Guillou, J.C. and Zinn-Justin, J. [1973] Phys. Rev. B8, 5330.

Bruce, A.D., Droz, M. and Aharony, A. [1974] J. Phys. C7, 3673.

Domb, C. and Green, M.S. (eds.)[to appear] Phase Transitions and Critical Phenomena, vol.6.

Ferrell, R.A. and Scalapino, D.J. [1972] Phys. Rev. Lett. 29, 413, [1972] Phys. Lett. 41A, 371.

Fisher, M.E. [1967] Rept. Prog. Phys. 30, 615.

Fisher, M.E. [1973] Nobel Symposium 24, 16.

Fisher, M.E. [1973] Phys. Rev. Lett. 30, 679.

Fisher, M.E. [1974] Rev. Mod. Phys. 46, 597.

Fisher, M.E., Ma, S. and Nickel, B.G. [1972] Phys. Rev. Lett. 29, 917.

Fisher, M.E. and Pfeuty, P. [1972] Phys. Rev. B6, 1889.

Golner, G.R. and Riedel, E.K. [1975] Phys. Rev. Lett. 34, 271, [1975] 34, 856.

Golner, G.R. and Riedel, E.K. [1975] Phys. Rev. Lett. 34, 856 and to be published.

Grover, M.K. [1972] Phys. Rev. B6, 3546.

Grover, M.K., Kadanoff, L.P. and Wegner, F.J. [1972] Phys. Rev. B6, 311.

Hikami, S. [1973] Prog. Theor. Phys. 49, 1096.

Ho, J.T. and Litster, J.D. [1969] J. Appl. Phys. 40, 1270.

Houghton, A. and Wegner, F.J. [1974] Phys. Rev. A10, 435.

Jasnow, D. and Wortis, M. [1968] Phys. Rev. 176, 739.

Kadanoff, L.P. [1966] Physics 2, 263.

Kadanoff. L.P. [1975] Phys. Rev. Lett. 34, 1005.

Kadanoff, L.P. et al [1967] Rev.Mod.Phys. 39, 395.

Kadanoff, L.P. and Houghton, A. [1975] Phys. Rev. B11, 377.

Ma, S. [1972] Phys. Rev. Lett. 42A, 5, [1973] Phys. Rev. A7, 2172.

Ma, S. [1973] Rev. Mod. Phys. 45, 589.

Mermin, N.D. and Wagner, H. [1966] Phys. Rev. Lett. 22, 1133.

Nauenberg, M. and Nienhuis, B. [1974] Phys. Rev. Lett. 33, 1598.

Niemeijer, Th. and van Leeuwen, J.M.J. [1973], Phys. Rev. Lett. 31, 1411, [1974] Physica 71, 17.

Rudnick, J. [1975] Phys Rev. Lett. 34, 438.

Shukla, P. and Green, M.S. [1974] Phys. Rev. Lett. 33, 1263, [1975] 34, 436.

Stanley, H.E. [1968] Phys. Rev. 176, 718.

Suzuki, M. [1972] Phys. Lett, 42A; [1973] Prog. Theor. Physik 49, 424.

Swift, J. and Grover, M.K. [1974] Phys. Rev. A9, 2579.

van der Waals, J.D. [1873] Doctoral dissertation. Leiden.

Wegner, F.J. [1972] Phys. Rev. B5, 4529.

Wegner, F.J. [1972] Phys. Rev. B6, 1891.

Wegner, F.J. [1973] Critical Phenomena and the Renormalization Group. An introduction (VIIIth Finnish Summer School in Theoretical Solid State Physics, August 1973, Siikajarvi, Finland).

Wegner, F.J. [1974] J. Physics C7, 2098.

Wegner, F.J. Phys Lett. [to appear].

Wegner, F.J. and Houghton, A. [1973] Phys. Rev. A8, 401.

Weiss, P. [1907] J. Physique 6, 661; [1911] Arch. Sc. Phys. et nat 31, 401.

Widom, B. [1965] J. Chem. Phys. 43, 3898.

Wilson, K.G. [1971] Phys. Rev. B4, 3184.

Wilson, K.G. [1971] Phys. Rev. B4, 3174.

Wilson, K.G. [1974] Physica 73, 119.

Wilson, K.G. and Fisher, M.E. [1972] Phys. Rev. Lett. 28, 240.

Wilson, K.G. and Kogut, J. [1974] Phys. Reports 12C, 75.

Zinn-Justin, J. [1973] Lectures given at the 1973 Cargese Summer School.

SCALING AND TIME REVERSAL FOR THE LINEAR MONOENERGETIC BOLTZMANN EQUATION

Erdal İnönü*

Middle East Technical University

ABSTRACT

For the linear, monoenergetic Boltzmann equation, we consider continuous and discrete scale transformations involving both the space-time parameters and the parameters of the given scattering kernel. One consequence is a correspondence between solutions of two problems which differ in the scattering kernel essentially by the addition of delta distributions in the forward and backward directions. This allows one to construct easily soluble models for extremely anisotropic scattering. There is also a similarity between the generalized scale group considered here and the renormalization group introduced for critical phenomena.

* Present address: Boğaziçi University, Bebek, Istanbul

1. INTRODUCTION

I wish to show here how some ideas taken from quantum mechanics can successfully be applied to an equation of classical statistical physics. The equation is the mono-energetic, linear Boltzmann equation and the main thought is to see whether a scale transformation leads to something non-trivial. We shall see that it indeed gives an interesting result, provided it is applied not only to space-time but also to the scattering kernel.

The linear, monoenergetic, homogeneous Boltzmann equation, used in the theories of neutron transport and radiative trans-fer can be written (in the terminology of neutron transport) as

$$\frac{1}{v}\ \frac{\partial\psi}{\partial t} + \vec{\Omega}.\vec{\nabla}\psi + \Sigma_t\psi(\vec{x},t,\vec{\Omega}) = \int\Sigma_s(\vec{\Omega}'\rightarrow\vec{\Omega})\ (\vec{x},t,\vec{\Omega}')d\vec{\Omega}' \qquad (1.1)$$

Here, $\psi(\vec{x},t,\Omega)$ is the single-particle distribution function giving the average number of neutrons around the phase space point $(\vec{x},\vec{\Omega})$ at time t. $\vec{\Omega}$ is the unit vector along the direction of motion of a neutron, v is the speed which is assumed to be the same for all neutrons $(\vec{v} = v\vec{\Omega})$. The equation has been linearized by neglecting the collisions bet-ween neutrons. However collisions between neutrons and fixed nuclei of the medium are taken into account. We assume that such a collision changes only the direction of motion of the neutron. The effect of these collisions is represented by the integral term where $\Sigma_s(\vec{\Omega}'\rightarrow\vec{\Omega})$ is the macroscopic differential cross-section for scattering from $\vec{\Omega}'$ into $\vec{\Omega}$ and the integral is taken over the whole sphere. The third term on the left represents the loss from the phase space point $(\vec{x},\vec{\Omega})$ due to absorbtion and scattering away from the direction $\vec{\Omega}$; Σ_t being the macroscopic total cross-section.

Defining the scattering probability distribution $f(\vec{\Omega}' \rightarrow \vec{\Omega})$ by:

$$\Sigma_s(\vec{\Omega}' \rightarrow \vec{\Omega}) = \Sigma_s \; f(\vec{\Omega}' \rightarrow \vec{\Omega}) \tag{1.2}$$

where Σ_s is the total scattering cross-section; introducing further, the average number of secondaries, c, as:

$$c = \frac{\Sigma_s}{\Sigma_t} \tag{1.3}$$

and measuring the distances in units of $\frac{1}{\Sigma_t}$, the equation is transformed into the following well-known form:

$$\frac{1}{v}\frac{\partial\psi}{\partial t} + \vec{\Omega}.\vec{\nabla}\psi + \psi(\vec{x},t,\vec{\Omega}) = c\int f(\vec{\Omega}' \rightarrow \vec{\Omega})\,\psi(\vec{x},t,\vec{\Omega}')\,d\vec{\Omega}' \tag{1.4}$$

Note that f is normalized:

$$\int f(\vec{\Omega}' \rightarrow \vec{\Omega})\,d\vec{\Omega} = 1 \;. \tag{1.5}$$

In order to study the behaviour of the equation under the scale transformations, it is more advantageous to combine the collision integral (the "gain") with the total scattering term (the "loss"). We define for this purpose a new kernel as:

$$g(\vec{\Omega}' \rightarrow \vec{\Omega}) = \delta(\vec{\Omega}' - \vec{\Omega}) - cf(\vec{\Omega}' \rightarrow \vec{\Omega}) \tag{1.6}$$

with the normalization

$$\int \delta(\vec{\Omega}' - \vec{\Omega})\, d\vec{\Omega} = 1.$$

The equation can in this way be transformed into the simpler looking form:

$$\frac{1}{v}\frac{\partial\psi}{\partial t} + \vec{\Omega}.\vec{\nabla}\psi + \int g(\vec{\Omega}' \rightarrow \vec{\Omega})\,\psi(\vec{x},t,\vec{\Omega}')\,d\vec{\Omega}' = 0., \qquad (1.7)$$

2. "NEW" SCALE TRANSFORMATIONS

Now, consider a scale transformation on space-time:

$$\tilde{\vec{x}} = \frac{\vec{x}}{\alpha} \;, \qquad \tilde{t} = \frac{t}{\alpha} \qquad (\tilde{\vec{v}} = \vec{v}) \qquad \text{where} \qquad \alpha > 0 \tag{2.1a}$$

Clearly, the equation (1.7) is not invariant under this scale change. However the equation will keep its form unchanged if we add to (2.1a) the transformation

$$\tilde{g}(\vec{\Omega}' \rightarrow \vec{\Omega}) = \alpha g(\vec{\Omega}' \rightarrow \vec{\Omega}). \tag{2.1b}$$

Therefore a solution of the equation with the kernel $\tilde{g}$ (call it $\tilde{\psi}$) may be obtained from the solution with the original kernel g simply by making the change of variables:

$$\tilde{\psi}(\tilde{\vec{x}}, \tilde{t}, \vec{\Omega}) \equiv \psi(\frac{\vec{x}}{\alpha}, \frac{t}{\alpha}, \vec{\Omega}) \tag{2.2}$$

What is the meaning of the transformation (2.1b)? Integrating with respect to $\vec{\Omega}$ over the whole sphere, we obtain:

$$1 - \tilde{c} = \alpha(1 - c). \tag{2.3}$$

while the correspondence (2.1b) can be more explicitly written as:

$$\tilde{c}\tilde{f}(\vec{\Omega}' \rightarrow \vec{\Omega}) = \alpha c f(\vec{\Omega}' \rightarrow \vec{\Omega}) + (1 - \alpha)\delta(\vec{\Omega}' - \vec{\Omega}) \tag{2.4}$$

Defining a new parameter,

$$k = \frac{1 - \alpha}{c} \qquad (\text{so that from (2.3): } 1 - k = \frac{\alpha c}{\tilde{c}}) \tag{2.5}$$

one can write (2.4) in a more intuitive form as:

$$\tilde{f}(\vec{\Omega}' \rightarrow \Omega) = (1 - k) f(\vec{\Omega}' \rightarrow \vec{\Omega}) + k\delta(\vec{\Omega}' - \vec{\Omega}). \qquad (2.6)$$

It is thus seen that the effect of adding an arbitrary delta distribution in the forward direction to the scattering kernel and diminishing at the same time the strength of the interaction so as to keep the total probability of scattering unchanged can be exactly taken care of by a simple change of scale in space-time.

One can use this correspondence to study the macroscopic effects of extreme forward anisotropy in the scattering kernel. A special advantage is that by changing k in eq.(2.6) continuously from zero to one, one can go over from the original f which may be taken to be completely isotropic, to the forward delta distribution, and see how various transport parameters change along the way (cf. Eriş, İnönü, Öztunali, Üsseli [1974]).

To point out another application, we note that the transformations (2.1a, 2.1b) form a group (with $\alpha > 0$), isomorphic to the multiplicative group of positive real numbers. This scale group may be combined with the usual invariance group of the linear, monoenergetic Boltzmann equation (namely, the three dimensional Euclidean group plus time translations) and one can use the irreducible representations of this larger group in order to characterize the solutions of the equation (cf.İnönü, Zweifel [1974]).

It may be interesting to remark finally that the generalized scale group under consideration is also a kind of exact renormalization group (cf. Wilson and Kogut [1974]) for the linear, monoenergetic Boltzmann equation. One makes a scale change with (2.1a) which changes the equation; then a new scattering kernel is defined by (2.2b) and the original form of the equation is exactly recovered. In the mean time, the constant c which in a sense, measures the "strength of the interaction" is changed into $\tilde{c}$ by the relation (2.3). The fixed points of this transformation are $c = 1$ and $c = \infty$. $c = 1$ coresponds to the case of no absorbtion (assuming there is no multiplication).

On the other hand, we know that the neutron diffusion length ν_o acts as a correlation length in the theory and furthermore $\frac{1}{\nu_o^2}$ can be expressed as a power series in 1-c, (for $c < 1$),

$$\frac{1}{\nu_o^2} = 3(1 - c)(1 - f_1) + 0[(1 - c)^2] \tag{2.7}$$

with a corresponding series for $c > 1$. Here, f_1 is the first Legendre coefficient in the expansion of the probability distribution $f(\vec{\Omega}' \to \vec{\Omega})$ in terms of Legendre polynomials (f is assumed to have rotational invariance):

$$f(\vec{\Omega}' \to \vec{\Omega}) = f(\vec{\Omega}' . \vec{\Omega}) = \frac{1}{2\pi} \sum_{\ell=0}^{\infty} (\frac{2\ell+1}{2}) f_\ell P_\ell (\vec{\Omega}' . \vec{\Omega}) \tag{2.8}$$

with $f_o = 1$, from the normalization (1.5).

Now, considering $c = 1$ as a critical point where the correlation length becomes infinite, we see that the approach of ν_o to its critical value is governed by the exponents $\nu = \nu' = 0.5$, corresponding to their values for a mean field theory (cf. Stanley [1971]).

One can find other similarities with characteristics of critical phenomena around the point $c = 1$, but we shall not pursue this matter further here.

3. INVERSIONS. TIME REVERSAL IN PARTICULAR

Let us now remove the restriction $\alpha > 0$ and consider in particular the inversions.

a) Under the transformation,

$$\tilde{\vec{x}} = -\vec{x}\,, \qquad \tilde{t} = -t \qquad (\tilde{\vec{\Omega}}' = \vec{\Omega}) \tag{3.1a}$$

$$\tilde{g} = -g \tag{3.1b}$$

the equation remains invariant. (This is somewhat reminiscent of the CPT theorem). However the correspondence (3.1b) leads to a physically non-realizable situation (f becoming negative), unless $f = \delta(\vec{\Omega}' - \vec{\Omega})$.

ii) A more interesting and useful case is obtained by considering the effect of time reversal on the stationary solutions, with the assumption of rotational symmetry for the scattering kernel. Our equation is now simply,

$$\vec{\Omega}.\vec{\nabla}\psi + \int g(\vec{\Omega}'.\vec{\Omega})\,\psi(\vec{x},\vec{\Omega}')\,d\vec{\Omega}' = 0 \tag{3.2}$$

where $g(\vec{\Omega}' \rightarrow \vec{\Omega}) = g(\vec{\Omega}'.\vec{\Omega})$ because of rotational invariance.

Under time reversal,

$$\tilde{\vec{x}} = \vec{x}\,, \qquad \tilde{t} = -t \longrightarrow \tilde{\vec{\Omega}} = -\vec{\Omega} \tag{3.3a}$$

and

$$\tilde{g} = -g \tag{3.3b}$$

the equation again keeps its form, since we have

$$\int g(-\vec{\Omega}'.\vec{\Omega})\psi(\vec{x},\vec{\Omega}')d\vec{\Omega}' = \int g(\vec{\Omega}'.\vec{\Omega})\psi(\vec{x},-\vec{\Omega}')d\vec{\Omega}' \quad . \tag{3.4}$$

In other words, if $\psi(\vec{x},\vec{\Omega})$ is the solution of eq.(3.2) with g, $\tilde{\psi}(\vec{x},\vec{\Omega}) = \psi(\vec{x},-\vec{\Omega})$ will satisfy the same equation with -g. Now, because of the averaging property expressed by eq. (3.4), it is possible to form out of $\psi(\vec{x},\vec{\Omega})$ and $\psi(\vec{x},-\vec{\Omega})$ a new solution of the equation (3.2) in the following way.

Introducing the notation

$$\psi(\vec{x},\vec{\Omega}) = \psi_+ \quad , \qquad \psi(\vec{x},-\vec{\Omega}) = \psi_- \tag{3.5}$$

and defining the matrices

$$\Psi(\vec{x},\vec{\Omega}) = \begin{pmatrix} \psi_+ \\ \psi_- \end{pmatrix} \quad , \qquad G(\vec{\Omega}'.\vec{\Omega}) = \begin{pmatrix} g(\vec{\Omega}'.\vec{\Omega}) & 0 \\ 0 & -g(\vec{\Omega}'.\vec{\Omega}) \end{pmatrix} \tag{3.6}$$

we can combine the equations satisfied by $\psi(\vec{x},\vec{\Omega})$ and $\psi(\vec{x},-\vec{\Omega})$ into the matrix equation:

$$\vec{\Omega}.\vec{\nabla}\Psi + \int G(\vec{\Omega}'.\vec{\Omega})\Psi(\vec{x},\vec{\Omega}')d\vec{\Omega}' = 0 \tag{3.7}$$

Consider a linear transformation of Ψ by the non-singular real 2×2 matrix,

$$L = \begin{pmatrix} A & B \\ C & D \end{pmatrix} \quad \text{with} \quad AD - BC \neq 0 \tag{3.8}$$

and look for the conditions that

$$\tilde{\Psi} = L\,\Psi \tag{3.9}$$

still gives solutions of the Boltzmann equation.

What are these conditions? First, since

$$\Psi(\vec{x},-\vec{\Omega}) = \sigma_x \Psi(\vec{x},\vec{\Omega}) \tag{3.10}$$

where $\sigma_x = \begin{pmatrix} 0 & 1 \\ 1 & 0 \end{pmatrix}$, L must commute with σ_x. It follows that L must have the form:

$$L = \begin{pmatrix} A & B \\ B & A \end{pmatrix} \quad \text{with} \quad A^2 - B^2 \neq 0 \tag{3.11}$$

Secondly, it must be possible to write the integral term as:

$$\int LGL^{-1}\,\tilde{\Psi}(\vec{x},\vec{\Omega}')\,d\vec{\Omega}' = \int \tilde{G}(\vec{\Omega}'.\vec{\Omega})\,\tilde{\Psi}(\vec{x},\vec{\Omega}')\,d\vec{\Omega}'$$

where

$$\tilde{G} = \begin{pmatrix} \tilde{g} & 0 \\ 0 & -\tilde{g} \end{pmatrix} \tag{3.12}$$

It turns out that because of rotational invariance and the property (3.4), this second condition is always satisfied once L is taken as in eq. (3.11).

Indeed, transforming the equation (3.7) with L, we obtain:

$$\vec{\Omega}.\vec{\nabla}\tilde{\Psi} + \int \tilde{G}(\vec{\Omega}'.\vec{\Omega})\,\tilde{\Psi}(\vec{x},\vec{\Omega}')\,d\vec{\Omega}' = 0 \tag{3.13}$$

where G has the form (3.12) with

$$\tilde{g}(\vec{\Omega}'.\vec{\Omega}) = \left(\frac{A^2+B^2}{A^2-B^2}\right) \cdot g(\vec{\Omega}'.\vec{\Omega}) - \left(\frac{2AB}{A^2-B^2}\right) \cdot g(-\vec{\Omega}'.\vec{\Omega}) \tag{3.14}$$

Therefore, if $\psi(\vec{x},\vec{\Omega})$ is a solution of the stationary equation (3.2) with g as the scattering kernel,

$$\tilde{\psi}(\vec{x},\vec{\Omega}) = A\psi(\vec{x},\vec{\Omega}) + B\psi(\vec{x},-\vec{\Omega}) \tag{3.15}$$

will be a solution of the same equation with $\tilde{g}$ as the kernel given by eq.(3.14).

Note that we have the relation:

$$\left(\frac{A^2+B^2}{A^2-B^2}\right)^2 - \left(\frac{2AB}{A^2-B^2}\right)^2 = 1 \tag{3.16}$$

Thus, the transformation of g depends on only one parameter. Introducing

$$\frac{A^2+B^2}{A^2-B^2} = \cosh K \qquad \frac{2AB}{A^2-B^2} = \sinh K \tag{3.17}$$

and the definitions

$$g(\vec{\Omega}'.\vec{\Omega}) = g$$

$$g(-\vec{\Omega}'.\vec{\Omega}) = g_-$$

we can write eq. (3.14) as:

$$\begin{aligned}\tilde{g} &= g \cosh K - g_- \sinh K \\ \tilde{g}_- &= -g \sinh K + g_- \cosh K\end{aligned} \tag{3.18}$$

and see that the transformation (3.14) for $\begin{pmatrix} g \\ g_- \end{pmatrix}$ is actually a Lorentz transformation in 1+1 dimensions. K varies over the whole line $-\infty < K < +\infty$ and determines the ratio of

A and B as :

$$\tanh \frac{K}{2} = \frac{B}{A} . \tag{3.19}$$

To make the physical meaning more clear, we go back to the scattering kernels f, $\tilde{f}$ and obtain from eq.(3.14):

$$\begin{aligned}\tilde{f}(\vec{\Omega}'.\vec{\Omega}) &= \frac{c \cosh K}{\tilde{c}} f(\vec{\Omega}'.\vec{\Omega}) - \frac{c \sinh K}{\tilde{c}} f(-\vec{\Omega}'.\vec{\Omega}) \\ &+ \frac{1-\cosh K}{\tilde{c}} \delta(\vec{\Omega}'-\vec{\Omega}) + \frac{\sinh K}{\tilde{c}} \delta(\vec{\Omega}'+\vec{\Omega})\end{aligned} \tag{3.20}$$

with

$$\tilde{c} - 1 = e^{-K}(c-1) \tag{3.21}$$

Defining new parameters k_1, k_2 as

$$k_1 = \frac{1-\cosh K}{\tilde{c}} , \qquad k_2 = \frac{\sinh K}{\tilde{c}} \tag{3.22}$$

(3.20) can be written as:

$$\begin{aligned}\tilde{f}(\vec{\Omega}'.\vec{\Omega}) &= e^{K}(1-k_1-k_2)[f(\vec{\Omega}'.\vec{\Omega})\cosh K - f(-\vec{\Omega}'.\vec{\Omega})\sinh K] \\ &+ k_1\delta(\vec{\Omega}'-\vec{\Omega}) + k_2\delta(\vec{\Omega}' + \vec{\Omega})\end{aligned} \tag{3.23}$$

In the case of a kernel f even in $(\vec{\Omega}'.\vec{\Omega})$, the relation simplifies to:

$$\tilde{f}(\vec{\Omega}'.\vec{\Omega}) = (1-k_1-k_2)f(\vec{\Omega}'.\vec{\Omega}) + k_1\delta(\vec{\Omega}'-\vec{\Omega}) + k_2\delta(\vec{\Omega}'+\vec{\Omega}) . \tag{3.24}$$

Thus, the combination of time reversal with rotational invariance allowed us to establish a correspondence between problems which differ essentially by the addition of a backward and a forward delta distribution in the scattering kernel.

Note that in eqs.(3.23 and 3.24), the parameters k_1, k_2 are not independent, as they satisfy the relation:

$$(1 - \tilde{c}k_1)^2 - (\tilde{c}k_2)^2 = 1 \tag{3.25}$$

It is however possible to apply the transformations (2.1b) and (3.14) in succession and obtain correspondences similar to eqs.(3.23-3.24) where k_1 and k_2 will be independent. This leads to an interesting theorem in anisotropic scattering which was previously derived by directly solving the transport equation. (cf. Lathrop [1965], İnönü [1973]).

REFERENCES

Case, K.M. and Zweifel, P.F. [1967] *Linear Transport Theory,* Massachusetts: Addison Wesley Pub. Co. Reading.

Eriş, A., Inönü, E., Öztunali, O. and Usseli, I. [1974] Models for Anisotropic Scattering in Neutron Transport Theory. IAEA Rep. 1228/RE.

Inönü, E. and Zweifel, P.F. [1974] *Proc. 3rd Int. Col. on Group Theoretical Methods in Phyics,* CNRS, vol.2, 351, Marseille.

Inönü, E. [1973] Transp. Theory and Stat. Phys. 3, 137.

Lathrop, K.D. [1965] Nuc. Sci. and Eng. 21, 498.

Stanley, H.E. [1971] *Introduction to Phase Transitions and Critical Phenomena,* Oxford University Press.

Wilson, K. and Kogut, J. [1974] Phys. Rep. 12c, 76.

SYSTEMS OF INFINITELY MANY DEGREES OF FREEDOM

Klaus Hepp

Physics Department, ETH
CH-8049 Zurich, Schweiz

1. SOME EXOTIC PROPERTIES OF INFINITE QUANTUM SYSTEMS

Systems of infinitely many degrees of freedom are frequently encountered in theoretical physics: in field theories and, asymptotically, in lattice and particle systems, both classical and quantum mechanical. A unified treatment of all interesting examples is clearly impossible, since there are infinitely many different ways of approaching the infinite as a limit from finite approximations. In this first lecture we shall just play on the beautiful shore of Bogazici and look at some exactly soluble quantum systems of infinitely many degrees of freedom which show a wide variety of exotic behavior.

Amplifiers: Properly built infinite systems can amplify small effect of $O(1)$ into an avalanche of $O(N)$, if the number N of degrees of freedom goes to infinity. A beautiful quantum mechanical amplifier has been invented by S. Coleman:

$$H_{(N)} = H_o + V_{(N)} , \qquad H_o = p = \frac{1}{i}\frac{\partial}{\partial x} ,$$

$$V_{(N)} = (\tfrac{1}{2} - S_o^3) \sum_{n=1}^{N} V(x-n) S_n^{'} , \tag{1.1}$$

$$V(x) = V(x)^* \in C_o^o(\mathbb{R}), \quad \int_{-\infty}^{+\infty} V(x)\, dx = \pi$$

This Hamiltonian describes the one-dimensional motion of an electron with spin $\frac{1}{2}$ and kinetic energy p along a chain of N spin $\frac{1}{2}$ systems. The spin operators S_m^i, $m = 0,1,2,\ldots$ satisfy $[S_m^i, S_n^j] = i\delta_{mn}\,\varepsilon^{ijk} S_m^k$. If the 3-component of the electron spin is positive, i.e. $S_o^3\, \varphi_o^+ = \frac{1}{2}\, \varphi_o^+$, then nothing happens except for a free translation. If the electron spin is $\varphi_o^- = 2S_o^3 \varphi_o^-$, then the interaction $V_{(N)}$ can flip $O(N)$ lattice spins. For instance, if the initial state was $\bigotimes_{n=1}^{N} \varphi_n^+$ then for large t one obtains essentially $\bigotimes_{n=1}^{N} \varphi_n^-$. This system can serve as a macroscopic pointer for a measurement of first kind in a Stern-Gerlach type experiment (Bell [1975], Berezin [1964]).

While the interaction picture time evolution $\exp(iH_o t)\exp(-iH_{(N)} t)$ of eq.(1.1) has a limit for $t \to +\infty$, even if $N = \infty$, there is a more catastrophic amplification effect in the Dicke maser model (cf. Dicke [1954])

$$H_{(N)}^S = a^* a + \varepsilon \sum_{n=1}^{N} S_n^3 + \frac{\lambda}{\sqrt{N}} \sum_{n=1}^{N} S_n^+ a + a^* S_n^-) \tag{1.2}$$

Here photons of frequency 1 are coupled to N 2-level atoms with energy difference $\varepsilon > 0$. G. Scharf has shown (cf. Scharf [1974]) that if, initially all atoms are excited in the state $\otimes\varphi_n^+$, then the maximum of the expectation value of the photon number $\gamma(t,M,N)$ depends very sensitively on the number M of photons initially present. For $\varepsilon = 1$ (resonance) one obtains for large N

$$\max_t \gamma(t,M,N) \simeq M + \frac{4M+4}{4M+5} N \qquad (1.3)$$

Again, an $0(1)$ initial difference is amplified into $0(N)$, but here in a rather explosive way. In the limit $N \to \infty$ one obtains the simple result (cf. Hepp [1973])

$$\gamma(t,M,\infty) = \frac{\lambda^2}{\rho^2} \operatorname{sh}^2\rho t + M\{\operatorname{ch}^2\rho t + \frac{(1-\varepsilon)^2}{4\rho^2} \operatorname{sh}^2\rho t\}$$

$$\rho^2 = \lambda^2 - (1-\varepsilon)^2/4 \qquad (1.4)$$

For not too large detuning, $(1-\varepsilon)^2 < 4\lambda^2$, one has now an unlimited, highly discriminating amplification.

Oscillators: Infinite weakly coupled systems can produce an alternating current as response to a time-independent voltage V. A simple model showing an a.c. Josephson effect (cf. Josephson [1962]) is

$$H_{(N)} = H^a_{(N)} + H^b_{(N)} + V_{(N)} \qquad (1.5)$$

$$H^a_{(N)} = \sum_{n=1}^{N} \varepsilon_a (a^*_{n+}a_{n+} + a^*_{n-}a_{n-}) - \frac{\lambda}{N} \sum_{m,n=1}^{N} a^*_{n+}a^*_{n-}a_{m-}a_{m+}$$

$$V_{(N)} = \frac{\tau}{N^2} \sum_{m,n=1}^{N} (a^*_{m+} a^*_{m-} b_{n-} b_{n+} + b^*_{n+} b^*_{n-} a_{m-} a_{m+})$$

Here the $a^{\#}_{n\pm}$ (resp. the $b^{\#}_{n\pm}$ in the similarly built $H^b_{(N)}$ with $\varepsilon_b = \varepsilon_a - eV$) are fermion creation and annihilation operators for the electron states of the a- (resp. b-) "superconductor" with attractive $(\lambda > 0)$ electron-electron interaction. $H^a_{(N)}$ and $H^b_{(N)}$ are $0(N)$ for $N \to \infty$, while the "weak link" $V_{(N)}$ is only $0(1)$. If one looks at states which lie close to the ground state $\Omega^a_{(N)} \otimes \Omega^b_{(N)}$ of $H^a_{(N)} + H^b_{(N)}$, each with occupation number ρN, $0 < \rho < 2$, then one obtains in the limit $N \to \infty$ a simple phase oscillator (cf.Hepp [1975]):

$$H_{(N)} \to H = \eta_a N_a + \eta_b N_b + \upsilon \cos(\phi_a - \phi_b), \qquad (1.6)$$

where $\eta_a = 2\varepsilon_a + \lambda(\rho-1) = \eta_b + 2eV$, $\upsilon = \tau(2\rho-\rho^2)$, and where the N's and ϕ's are quantum mechanical action and angle operators (cf. Carruther and Nieto [1968]) with

$$[\phi_a, N_a] = [\phi_b, N_b] = i. \qquad (1.7)$$

Therefore the electron current from a to b is harmonic:

$$J_{a \to b}(t) = -e \frac{d}{dt} e^{iHt} N_a e^{-iHt} =$$
$$(1.8)$$
$$= -e\upsilon \sin(\phi_a(t) - \phi_b(t) = -e\upsilon \sin(\phi_a - \phi_b + 2eVt)$$

Solitons: Infinite spin systems can exhibit in their ground state representation an energy spectrum which can be resolved in terms of multiparticle states of elementary excitations. A well understood non-trivial example is the one-dimensional ferromagnetic Heisenberg chain

$$H_{(N)} = -\frac{1}{2} \sum_{n=1}^{N} \{S_n^+ S_{n+1}^- + S_{n+1}^- S_n^+ + 2\Delta(S_n^3 S_{n+1}^3 - \frac{1}{4})\} \qquad (1.9)$$

with anisotropy $\Delta = \text{ch}\ \delta > 1$ and periodic boundary condition $S_{N+1}^i = S_1^i$, where $S_n^{\pm} = S_n^1 \pm i\, S_n^2$. One ground state is $\phi_{(N)} = \bigotimes \varphi_n^-$. The one-magnon eigenstates

$$\phi_{(N)}(k) = \sum_{n=1}^{N} e^{ikn} S_n^+ \phi_{(N)}, \qquad k = \frac{2\pi m}{N}, \ m=1,\ldots N \qquad (1.10)$$

are obvious eigenstates of $H_{(N)}$ in the invariant subspace $\mathcal{H}_{(N)}^L$ of $\Sigma(S_N^3 - \frac{1}{2}) = L$, for $L = 1$ spin deviations. H.A. Bethe [1931] has given a brilliant heuristic characterization of the spectrum and the eigenstates of $H_{(N)}$ for arbitrary L, which is the basis of the following rigorous results by S. Denoth [1975]: For sufficiently large N there exist families of states which, up to exponentially vanishing corrections, describe clusters $\{F_1,\ldots,F_f\}$ of fragments of F_i bound spin deviations, for all $F_i = 1,2,\ldots$ with $\Sigma F_i = L$. For $N = \infty$ these states become generalized eigenstates of $H_{(\infty)}$, which can be combined into scattering states with "soliton" behavior: incoming wave-packets with free fragments $\{F_1,\ldots, F_t\}$ for $t \to -\infty$ only lead to the same fragments and momenta for $t \to +\infty$, without coupling between different channels (as in multichannel scattering theory (cf. Faddeev [1963])) by disintegration or recombination of bound states. For $L = 2,3$ Denoth has proved asymptotic completeness. This one-dimensional spin chain is one of the rare examples of exactly soluble models (in the limit $N \to \infty$) which has short range interactions. In the limit (lattice constant) $= a \to 0$ and $\Delta = 1+a\lambda/2$ this model leads to L 1-dimensional bosons with attractive δ-interaction.

$$H^{L}_{(\infty)}(a) \to -\frac{1}{2}\sum_{\ell=1}^{L}\frac{\partial^2}{\partial x_\ell^2} - \lambda \sum_{1\le\ell<m\le L}\delta(x_\ell - x_m) \qquad (1.11)$$

where the exact solution is known (cf. Berezin, Pohil and Finkelberg [1964]).

Equilibrium phase transitions: The most intensively studied "pathologies" of infinite systems are the phase transitions in thermal equilibrium. The free energy per atom of the Dicke model (eq.1.2), $F_{(N)}(\beta) = -(N\beta)^{-1}\,\ell n\; T\; \exp - \beta H^S_{(N)}$ can be computed exactly in the limit $N \to \infty$ (cf. Hepp and Lieb [1973])

$$\lim_{N\to\infty} F_{(N)}(\beta) = \begin{cases} \lambda^2\sigma(\beta)^2 - \dfrac{\varepsilon^2}{4\lambda^2} - \beta^{-1}\ell n(2\mathrm{ch}(\beta\lambda^2\sigma(\beta))), & \beta > \beta_c \\ \\ -\beta^{-1}\;\ell n(2\mathrm{ch}(\frac{\beta\varepsilon}{2})) & \beta \le \beta_c \end{cases} \qquad (1.12)$$

The critical temperature $T_c = (k\beta_c)^{-1}$ is zero for $\varepsilon \geq \lambda^2$ and otherwise $\beta_c = \frac{2}{\varepsilon}$ arth (ε/λ^2), and $\sigma(\beta)$ is for $\beta > \beta_c$ the nontrivial solution of the "gap equation" $2\sigma(\beta)=\text{th}(\beta\lambda^2\sigma(\beta))$.

Nonequilibrium phase transitions: If one looks at the Dicke Hamiltonian $H^S_{(N)}$ as a caricature of a laser cavity, then the time evolution has to contain a coupling $H^R_{(N)}$ to the outside world, which describes photon losses and the pumping of the atoms into a mean inversion $-\frac{1}{2} \leq \eta \leq \frac{1}{2}$. There exist models (cf. Hepp and Lieb [1973]) for the reservoir $H^R_{(N)}$, where the macroscopic observables

$$\sigma^i_{(N)}(t) = N^{-1} \sum_{n=1}^{N} S^i_n(t),$$
$$\alpha_{(N)}(t) = N^{-1/2} a \tag{1.13}$$

(propagated under the quantum mechanical time-evolution $H^S_{(N)} + H^R_{(N)}$) have a limit for $N \to \infty$, where they satisfy ordinary dissipative differential equations. Nonequilibrium phase transitions in this class of models are nonanalytic changes of the attractors of classical orbits.

In resonance $(\varepsilon = 1)$ the 1-mode laser equations (on an attracting invariant manifold) are the Lorenz equations

$$\dot{x} = -fx + fy,$$
$$\dot{y} = -xy + gx - y, \tag{1.14}$$
$$\dot{z} = xy - hz,$$

which have been numerically studied by Lorenz [1963]. For certain values of the damping constants the flow of (1.14) can be globally controlled by Liapunov functions, and there is only one non-equilibrium transition from a non-radiating to a coherently radiating attractor. For other values (contrary to the claim in Hepp and Lieb [1973]), this monochromatic radiation pattern becomes unstable for higher pumping η and dissolves into a rather chaotic motion (cf. Lorenz [1963]). By taking more modes and more general couplings, one expects in such models to find "strange attractors" in the sense of Ruelle and Takens [1971] with a "turbulent" radiation field.

2. CHANCE AND NECESSITY

In the past decade, the theory of infinite systems has been most successful in elucidating the foundations of equilibrium statistical mechanics (cf. Ruelle [1969]) and kinetic theory (cf. Lanford [1975]). There is the deep philosophical question (cf. Grad [1967]) how to reconcile the deterministic (at least in the classical case) and time-reversal (or TCP) invariant laws of microphysics with the very successful irreversible and stochastic dynamical equations of macrophysics. Common sense imposes a "coarse-grained" description on systems with many degrees of freedom, since we can neither uniquely specify the initial conditions of some 10^{23} mass points evolving in our laboratory, nor can we compute their orbits from the laws of classical or quantum mechanics. In this respect the exotic models described in the last section were useful, but highly artificial abstractions of the real world, since none of the "real" dynamical systems has an exact analytic solution for $N \to \infty$.

Before we come to the infinite systems approach to the ensembles of thermoelasticity, (which can be generalized to quantum mechanics without major difficulties) we shall describe Boltzmann's ergodic approach for classical canonical systems

$$\dot{p}_n = -\frac{\partial H}{\partial q_n}, \quad \dot{q}_n = \frac{\partial H}{\partial p_n}, \qquad 1 \leq n \leq N$$

$$H = H(x|\Lambda, N) = \sum_{n=1}^{N} \frac{p_n^2}{2m} + \sum_{1 \leq m < n \leq N} \phi(q_m - q_n) + B(x|\Lambda, N) \tag{2.1}$$

Here $\phi(q) = \phi(-q)$ is a smooth finite range potential with a hard core of diameter d and $B(x|\Lambda,N) = \sum_{n=1}^{N} \psi(q_n|\Lambda)$ a boundary term, which retains the particles in the container Λ (e.g. a rectangular box $\Lambda(a_1,a_2,a_3)$ with edge length a_1,a_2,a_3 centered at o). In this case the canonical equations (2.1) have global solutions for all allowed initial conditions defining a flow T^t on phase space

$$M(\Lambda,N) = \{x \in \mathbb{R}^{6N} | q_i \in \Lambda \ , \ |q_i - q_j| \geq d \quad \forall\, i \neq j \} \tag{2.2}$$

as well as on every energy surface

$$\Sigma(E,\Lambda,N) = \{x \in M(\Lambda,N) | \ H(x|\Lambda,N) = E\}. \tag{2.3}$$

By Liouville's theorem the Lebesgue measure dx is invariant under T^t, as well as the microcanonical measure

$$\begin{aligned} \mu_m(dx|E,\Lambda,N) &= \delta(H(x|\Lambda,N)-E)\,dx/N!\ Z(E,\Lambda,N) \\ Z(E,\Lambda,N) &= \int \delta(H(x|\Lambda,N)-E)\,dx/N! \end{aligned} \tag{2.4}$$

on $\Sigma(E,\Lambda,N)$ which is a manifold for almost all energies E. (If grad $H(x|\Lambda,N)$ vanishes, then (2.4) can be defined by a limiting process (cf. Lanford [1975]). We shall always take regular values of E.

The central postulate of equilibrium statistical mechanics is that this closed N-particle system with energy E should for all "questions of thermostatics" be taken in the statistical state $\mu_m(dx|E,\Lambda,N)$. This means that the value of any "thermostatic observable,"

$$\bar{f}(x) = \lim_{T\to\infty} \frac{1}{T} \int_o^T dt\ f(x(t)), \qquad (2.5)$$

the time-average over a function f on $\Sigma(E,\Lambda,N)$, should be evaluated using the probability measure μ_m. Boltzmann has forcefully argued that thermostatics should be described by time averages, and Birkhoff has proved the existence of $\bar{f}(x)$ for almost all $x \in \Sigma(E,\Lambda,N)$, if $f \in L'(\Sigma(E,\Lambda,N))$. $\bar{f}$ is μ_m- measurable, T^t-invariant and

$$\mu_m(f) = \int f(x)\ \mu_m(dx) = \mu_m(\bar{f}) \qquad (2.6)$$

The dynamical system (Σ,μ_m,T^t) is called ergodic, if the only μ_m-measurable T^t-invariant sets on Σ are either null sets or complements of null sets. Since $\mu_m(\Sigma) = 1$, one can say equivalently that all T^t-invariant functions $f \in L'(\Sigma)$ are constants, or that for almost every x and for every measurable set $\Delta \subset \Sigma$ the average time which $x(t)$ spends in Δ is equal to $\mu_m(\Delta)$.

For ergodic systems probability is a consequence of dynamics in the following precise sense: One should only consider "imprecise initial data" of the system, given by probability measures $\mu_\rho(dx) = \rho(x)\mu_m(dx)$, where $0 \leq \rho \in L'(\Sigma)$ and $\mu_m(\rho) = 1$. Then for an ergodic system the time average of μ_ρ converges weakly to μ_m for $t \to \pm\infty$. Since $\mu_{\rho(t)}(f) = \mu_\rho(f(-t))$ (where $\rho(t)(x) = \rho(x(t))$ one sees that in an ergodic system the value of any thermostatic observable $\bar{f}$ in every state μ_ρ

satisfies

$$\mu_\rho(\bar{f}) = \mu_m(f) \tag{2.7}$$

The enormous practical importance of (2.7) is that the r.h.s. can be evaluated without knowing (except for ergodicity) the solutions of the equations (2.1).

Using the mean ergodic theorem

$$\lim_{T\to\infty} \mu_m((f^T-\bar{f})^2) = 0, \qquad f^T(x) = \frac{1}{T}\int_0^T dt\ f(x(t)), \tag{2.8}$$

one has in an ergodic system smallness of the fluctuations of f^T for large T:

$$\lim_{T\to\infty} \mu_m((f^T- \mu_m(f))^2) = 0 \tag{2.9}$$

Therefore the microcanonical probability of the deviation of the time average f^T from $\mu_m(f)$ goes to zero for $T \to \infty$.

Time-like ergodic theory would give to the physicist a fairly satisfactory justification for using the microcanonical ensemble in computing thermostatic observables in all macroscopically specified states. However, nature is unkind to us, since the KAM theorem (cf. Kolmogorov [1954], Arnold [1963], Moser [1966]) provides us with a class of non-ergodic systems which remain so under small smooth perturbations.

It is interesting that in the thermodynamic limit (i.e. for fixed energy density $\varepsilon = E/|\Lambda|$ and particle density $\rho = N/|\Lambda|$ and for $|\Lambda| \to \infty$) one can justify the ensembles of equilibrium statistical mechanics for completely different reasons using the space-like ergodicity of pure thermodynamic phases (cf. Lanford [1973], Ruelle [1965]).

The natural observables for a large system of identical particles are functions on phase space, which depend on the coordinates $x = (x_1, \ldots, x_N)$ in a symmetrical way, e.g. the cylinder functions

$$F(x) = \sum_{n_1,\ldots,n_k=1}^{N}{}' f(x_{m_1},\ldots,x_{m_k}) \tag{2.10}$$

Here Σ' extends only over distinct indices and $f \in C_0^0$ for convenience. Then $\mu_m(F)$ can be expressed using the micro-canonical correlation function

$$\mu_m(F) = \int dx_1 \ldots dx_k \; f(x_1,\ldots,x_k) \; \rho(x_1,\ldots,x_k|E,\Lambda,N) \tag{2.11}$$

$$\rho(x_1,\ldots,x_k|E,\Lambda,N) = \int dx_{k+1} dx_N \; \delta(H(x|\Lambda,N)-E)/(N-k)!Z$$

Suppose that for ε,ρ fixed and $\Lambda = \Lambda(a_1,a_2,a_3)$, $a_i \to \infty$, the measures $\mu^\wedge = \mu_m(E,\Lambda,N)$ converge on all cylinder functions to a probability measure μ on the set of all possible configurations of the infinite system. Suppose further that μ is space-translation invariant,

$$\mu(F) = \mu(F^a) \qquad \forall a \in \mathbb{R}^3 \tag{2.12}$$

and ergodic with respect to space-translations, where F^a corresponds to $f^a(x,\ldots,x_k) = f(q_1+a,p_1,\ldots,q_k+a,p_k)$. We shall call the space-average $F^\wedge = |\Lambda|^{-1}\int da\; F^a$ a "macroscopic observable". Then

$$\lim_{\Lambda\to\infty} \mu((F^{\wedge}-\mu(F))^2) = 0 \tag{2.13}$$

and by the Lebesgue theorem

$$\lim_{\Lambda\to\infty} \mu^{\wedge}((F^{\wedge}-\mu^{\wedge}(F^{\wedge}))^2) = 0 \tag{2.14}$$

Hence, whenever the microcanonical ensembles $\mu^{\wedge}$ have a space-like ergodic weak limit μ, then the macroscopic observables of the system converge to constants in probability on the energy surfaces $\Sigma(E,\Lambda,N)$ with the normalized Lebesgue measure as "apriori probability". A special class of macroscopic observables are for $N = \rho|\Lambda|$ the summatory functions

$$F^N(x_1,\ldots,x_\ell) = N^{-1} \sum_{n_1,\ldots,n_\ell=1}^{N}{}' f(p_{n_1},\ldots p_{n_k}, q_{n_1}-q_{n_2},\ldots q_{n_{k-1}}-q_{n_k}) \tag{2.15}$$

with $f \in C_o^o$.

This "geometric" property of the energy surfaces $\Sigma(E,\Lambda,N)$ explains why the macroscopic aspects of a large system are describable in terms of very few parameters, as ε and ρ for (2.1). The space-like ergodic limit μ is called a thermodynamic equilibrium state of density ρ and energy density ε for a pure phase. Physical experience tells us that for values of ρ,ε, where two phases, like vapor and water, are in equilibrium, the space-average over the particle number should fluctuate, and this can be rigorously verified in some models (cf. Minlos and Sinai [1967], Minlos and Sinai [1968]).

For classical particle systems with a Hamiltonian (2.1) the existence of the limit $\mu = \lim \mu^{\wedge}$ and its space-like ergodicity can be shown for small ρ and large ε, or more physically for small activity and high temperatures (cf.Ruelle [1965]). In this case $0(N)$-fluctuations of macroscopic observables are exponentially small (cf. Lanford [1973]), while $O(\sqrt{N})$-fluctuations are Gaussian distributed. For the two-dimensional Ising model (cf. Gallavotti [1972]), much more complete results are known about the existence and ergodicity properties of the space-average of the classical ensembles (cf. Messager and Miracle-Sole [1975]) and the central limit theorem (cf. Martin-Löf [1973], Alfina and Minlos [1970]) away from the critical point. The interesting statistical properties at the critical point (cf. Wegner [1976]) have been investigated by Bleher and Sinai (cf. Bell [1975]) for the "hierarchical model".

While in the pure phase region of ρ,ε the use of the microcanonical ensemble can be founded on purely geometric arguments, one has to investigate the detailed properties of the flow T^t in a justification of kinetic theory. This program has been recently pushed ahead by O.E. Lanford [1975].

Consider a system of N hard spheres of diameter d in a box Λ. The 1-particle distribution function $f_1(t,x) \geq 0$, $\int f_1(t,x) = 1$, satisfies in kinetic theory the Boltzmann equation

$$\frac{\partial}{\partial t} f_1(t,x_1) = -\frac{p_1}{m}\frac{\partial}{\partial q_1} f_1(t,x_1)$$

$$+c \int_{\{\omega\cdot(p_2-p_1)\geq 0\}} d\omega dp_2 \frac{\omega\cdot(p_2-p_1)}{m}\{f_1(t,p_1',q_1)f_1(t,p_2',q_1)- \tag{2.16}$$

$$-f_1(t,p_1,q_1)f_1(t,p_2,q_1)\}$$

Here c is a constant and the ω-integration goes over all unit vectors in $\mathbb{R}^3$ with $\omega\cdot(p_2-p_1) \geq 0$, and p_1', p_2' are outgoing momenta in

$$p_1' = p_1 + [\omega\cdot(p_2-p_1)]\omega ,$$
$$\tag{2.17}$$
$$p_2' = p_2 + [\omega\cdot(p_1-p_2)]\omega .$$

In classical mechanics, the rescaled time-dependent j-particle correlation functions $f_j(t,x_1,\ldots,x_j)$ are derived from the measure $\mu(t,x_1,\ldots x_N)dx = \mu(x(-t))dx$ of a macroscopically specified state via

$$f_j(t,x_1\ldots x_j) = \frac{N!}{N^j(N-j)!}\int dx_{j+1}\ldots dx_N \mu(t,x_1\ldots x_N) \tag{2.18}$$

They satisfy the BBGKY hierarchy, which for $f_1(t,x_1)$ is

$$\frac{\partial}{\partial t} f_1(t,x_1) = -\frac{p_1}{m}\frac{\partial}{\partial q_1} f_1(t,x_1)$$

$$+ dN^2\int_{\{\omega(p_2-p_1)\geq 0\}} d\omega dp_2\{f_2(t,p_1',q_1,p_2',q_1-d\omega)-f_2(t,p_1,q_1,p_2,q_1+d\omega)\}.$$

It is tempting to perform the Boltzmann-Grad limit (cf. Grad

[1958]):

$$d \to 0, \qquad N \to \infty, \qquad c = d^2 N \quad \text{fixed}, \tag{2.20}$$

and to believe in "molecular chaos"

$$f_2(t,p_1,q_1,p_2,q_2) = f_1(t,p_1,q_1)f(t,p_2,q_2) \tag{2.21}$$

so that the Boltzmann equation should become the exact limiting dynamics of an infinite particle system of vanishing diameter with fixed mean free path.

The first result of Lanford is a purely geometrical justification of the assumption of "molecular chaos" for a representative ensemble of N hard spheres in Λ with diameter d in the limit (2.20): The macroscopic specification of configurations of kinetic theory (replacing ρ, ε) are continuous $f \geq 0$, $\int f dx = 1$. A reasonable approximation for finite N is obtained by choosing for every (N,d) with (2.20) an integer K, K non-intersecting subsets $\Delta_1, \ldots \Delta_K$ of $\Lambda \times \mathbb{R}^3$ of finite volume (satisfying $\lim_{N\to\infty} \inf_{1\leq i\leq K} N|\Delta_i| = \infty$ and another mild regularity condition) and K non-negative integers $N_1, \ldots N_K$ with $\Sigma N_i = N$ such that

$$\bar{f}^d(x) = \begin{cases} \dfrac{N_i}{N|\Delta_i|}, & \text{if} \quad x \in \Delta_i \\ 0, & \text{if} \quad x \notin \cup\Delta_i \end{cases} \tag{2.22}$$

approximates $f(x)$ uniformly for $N \to \infty$ with

$$\lim_{d\to 0} \sup_x \bar{f}^d(x)\, e^{\beta q^2} \leq Z' < \infty \tag{2.23}$$

for some $Z' < \infty$.

Let $\mu^{(d)}(x)dx$ be the probability measure in the phase space (2.2) of N hard spheres obtained by restricting the Lebesgue measure to those phase points with exactly N_i particles in Δ_i, $1 \leq i \leq K$. This is the kinetic analogue to some $\hat{\mu}$ in the equilibrium theory. Then it can be shown that "molecular chaos" holds asymptotically: the rescaled correlation functions $f_j^{(d)}(x_1,\ldots, x_j)$ (2.18) converge uniformly on compact sets in $(\Lambda_j \times \mathbf{R}^3)^j \setminus \{x \mid q_i = q_j \text{ for some } i \neq j\}$ to $\pi_{i=1}^j f(x_i)$ and for any $Z > Z'$

$$\sup_{j,x} (Z^{-j} \exp \beta \sum_{i=1}^j q_i^2)\, f_j^{(d)}(x_1,\ldots, x_j) < \infty. \tag{2.24}$$

Using time-dependent perturbation theory in the BBGKY hierarchy and the estimate (2.24), Lanford then proves the existence of the limit (2.20) in the solutions $f_j^{(d)}(t,x_1,\ldots, x_j)$ of the BBGKY hierarchy with initial condition $f_j^{(d)}(x_1,\ldots, x_j)$ at $t = 0$, for all sufficiently small positive times:

$$\lim_{d\to 0} f_j^{(d)}(t,x_1,\ldots, x_j) = \prod_{i=1}^j f(x_i,t) \tag{2.25}$$

exists almost everywhere for all j and $f(x,t)$ is a (generalized) solution to the Boltzmann equation (i.e. a solution to the "Boltzmann hierarchy") with initial condition $f(x)$.

Hence in so far as $\mu^{(d)}(x_1, \ldots, x_N)dx$ describes a representative kinetic ensemble for the hard sphere system, matching the macroscopic data $f(x)$, then the solution $f(t,x)$ of the Boltzmann equation describes the time-evolution of this typical ensemble. This beautiful rigorous result has been extended by F. King [1975] to soft potentials.

We are now much closer in the dynamical understanding of the relation between chance and necessity.

REFERENCES

Alfina, A. and Minlos, R. [1970] Izv. Akad. Nauk. SSR 34, 1173.

Arnold, V.I. [1963] Usp. Mat. Nauk 18, 91.

Bell, J.S. [1975] Helv. Phys. Acta 48, 99.

Berezin, F.A., Pohil, G.P. and Finkelberg, V.M. [1964] Vestnik Mosk. Univ. 1, 21.

Bethe, H.A. [1931] Z. Phys. 71, 205.

Bleher, P.M. and Sinai, Ya. G. [1973] Comm. Math. Phys. 33, 23. and to appear.

Carruthers, P. and Nieto, M.M. [1968] Rev. Mod. Phys. 40, 411.

Denoth, S. [1975] Thesis ETH Zurich.

Dicke, R.H. [1954] Phys. Rev. 93, 99.

Faddeev, L.D. [1963] Trudy Mat. Inst. Steklov 69

Gallavotti, G. [1972] Riv. Nuovo Cim. 2, 133.

Grad, H. [1958] in "Handbuch der Physik, vol.12" Berlin: Springer.

Grad, H. [1967] in "Delaware Seminar in the Foundations of Physics" (M. Bunge ed.) Berlin: Springer.

Hepp, K. [1972] Helv. Phys. Acta 45, 237.

Hepp, K. and Lieb, E.H. [1973] Ann. Phys. 76, 360.

Hepp, K. and Lieb, E.H. [1973] Helv. Phys. Acta 46, 573.

Hepp, K. [1975] Ann. Phys. 90, 285.

Josephson, B. [1962] Phys. Lett. 1, 251.

King, F. [1975] Thesis Univ. California, Berkeley.

Kolmogorov, A.N. [1954] Proc. Int. Cong. Math. Amsterdam.

Lanford, O.E. [1973] in "Statistical Mechanics and Mathematical Problems" (A. Lenard ed.) Berlin: Springer.

Lanford, O.E. [1975] in "Dynamical Systems, Theory and Applications" (J. Moser ed.) Berlin: Springer.

Lorenz, E.N. [1963] J. Atmos. Sci. 20, 130.

Martin-Löf, A. [1973] Comm. Math. Phys. 32, 75.

Messager, A. and Miracle-Sole, S. [1975] Comm. Math. Phys. 40, 187.

Minlos, R.A. and Sinai, Ya.G. [1967] Math USSR-Sbornik 2, 355.

Minlos, R.A. and Sinai, Ya.G. [1968] Trans. Moscow Math. Soc. 19, 121.

Moser, J. [1966] Ann. Scuola Norm. Sup. Pisa 20, 266 & 499.

Ruelle, D. [1965] J. Math. Phys. 6, 201.

Ruelle, D. [1969] "Statistical Mechanics" New York: Benjamin.

Ruelle, D. and Takens, F. [1971] Comm. Math. Phys. 20, 167.

Scharf, G. [1974] Ann. Phys. 83, 71.

Wegner, F.J. [1976] this volume.

BEYOND SYMMETRY:
HOMEOMETRY AND HOMEOMETRY GROUPS

A. O. Barut

Department of Physics,
The University of Colorado
Boulder, Colorado 80309

"Mathematics is the art of giving
the same name to different things"

H. Poincaré

Much has been written on the concept of symmetry in general, on symmetry groups and on symmetry principles in physics, in particular. (cf. Park [1968]). We shall not dwell here on symmetry, but try to go *beyond symmetry*. There is a whole *terra incognita* between symmetry and complete chaos, amorphy:

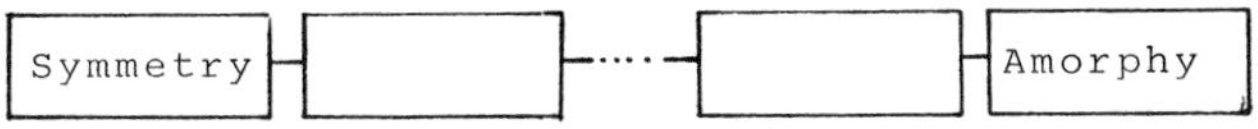

The unknown territory between symmetry and amorphy must be studied from geometric point of view and new structures must be identified and classified.

In physics, anything that is not symmetric is attributed, at first summarily, to the interplay of complex dynamical interactions. In some cases one speaks of "approximate" or "broken" symmetries, approximate because, again, of dynamical complications. But "broken symmetry" is a negative concept, implies imperfection, and the emphasis is on the symmetry. However, there are precise new structures which are not symmetric in the narrow sense, but could still be viewed as perfect. Consider the shapes shown in Fig. 1. They are not symmetric in the sense of Euclidean geometry, but have a fundamental elementary quality. In a more qualitative interpretation of the concept of symmetry, the word *harmony* may encompass this new situation. But the problem for physics and mathematics is to render this general concept of harmony more precise and quantitative. Symmetry groups perform this task for geometry, and for some theories of physics.

Amorphy, in a general sense, is the state of phenomena without definite shape, form, character, organization, purpose, and direction, limits and structure; we view these phenomena to be *unclassifiable*, in which cause and effect and controlling influences are not discernable. In more physical terms, such phenomena do not seem to be describable by *a finite number of physical laws*.

Between the two extremes of symmetry and amorphy we can imagine various intermediate levels in which the variables undergo transformations, which in the case of shapes and figures are not Euclidean transformations. For example, in Fig. 2, the number of prongs, the size, angles and distances of each set of prongs or any other quality is varied along the line. In Fig.3, some simple topologically equivalent variations of the circle

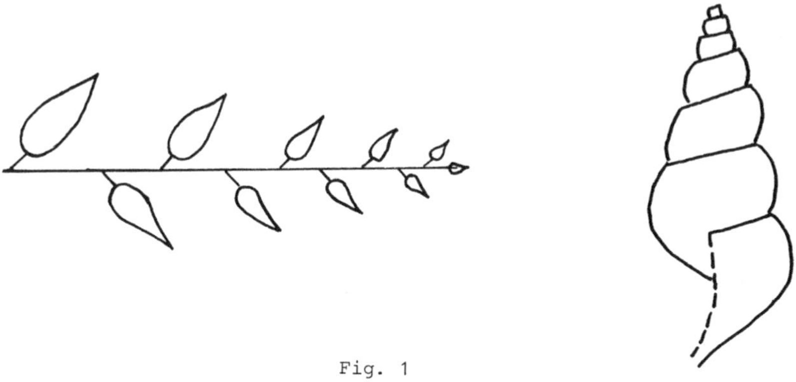

Fig. 1

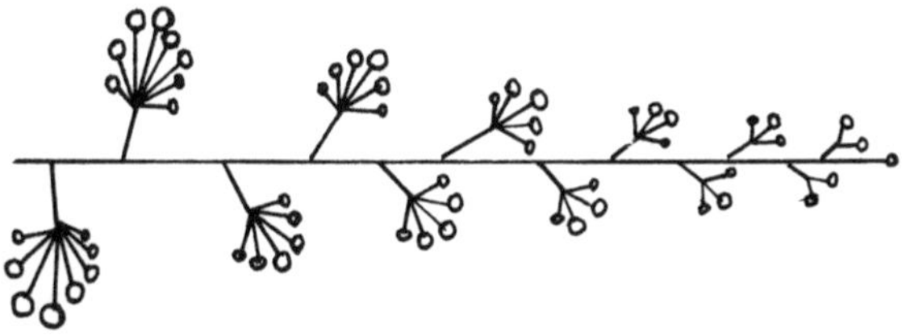

Fig. 2

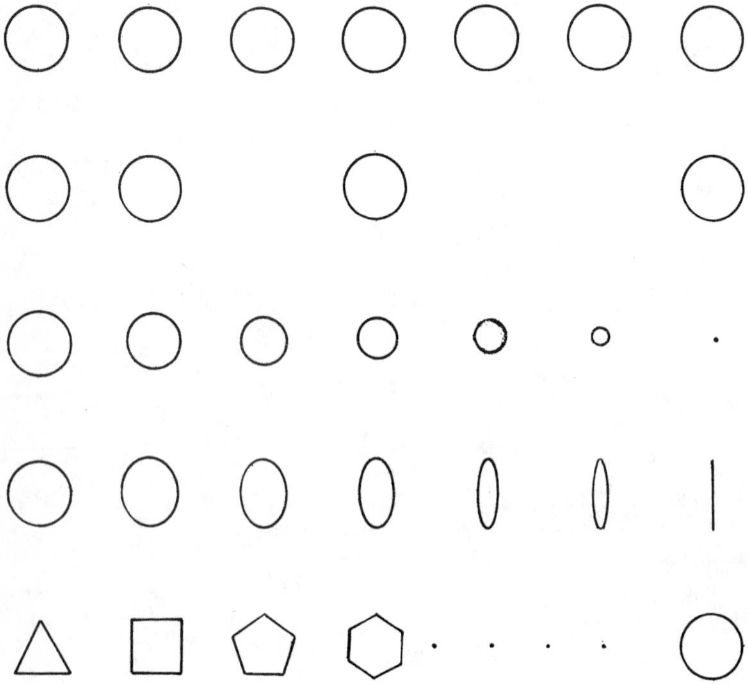

Fig. 3

are shown, in each line only a single quality being varied. It is easy to combine these transformations, or to imagine more complicated ones.

Definition of Homeometry. We define *homeometry* as the state of phenomena in which we recognize, in addition to the geometrical symmetry transformations, new principles of change of other variables or qualities of a basic system or form.

A more precise mathematical discussion will be given below.

The terminology expressing the relationship of forms ranges from automorph, isomorph, via homeomorph, syngemorph (inborn, related), katamorph (akin to), to heteromorph (different forms at different times and places, variable shape) and finally to amorph. We have selected the notion of *homeometry* to emphasize the common measure; *heterometry* may be used if the *change* is to be emphasized, rather than the common origin upon which (small) changes are superimposed.

Homeometry transformations "break" the symmetry under the geometrical transformations. But they provide a larger group of exact symmetry transformations if we adjoin to the geometrical operations also those of homeometry. For example, in Fig. 1, we have no longer translational invariance, but the product of translational and a certain scale transformation. By decomposing the new situation into translational and scale transformations we thus recognize the larger symmetry underlying a seemingly non-symmetrical situation.

It is clear that this process aims at reducing dynamics to geometry, by taking account properly of the changes of properties and qualities of the system from place to place, and in time. More general group of transformations which include part of the dynamics have been called dynamical groups. (For more recent work including material on dynamical groups see Englefield [1972], Wybourne [1974], Hermann [1966], Gilmore [1974], Barut [1972], Miller [1973], Loeble [1975], Moshinsky [1968], Barut and Raczka [1977]). Those homeometry groups include and generalize the concept of dynamical groups.

The symmetry principles (and as we shall see, also the homeometry principles) are used in physics via *the idea of relativity* as expressed in F. Klein's Erlangen program [1893] in 1872, in the case of Euclidean geometry. According to Klein, the points of the Euclidean space are indistinguishable, but their coordinates are "man-made"; objective properties of points must be independent of coordinate frames, and all cartesian coordinates are equally admissible. Now the coordinates of a point in different frames are related by a group of transformations. We can also say that a transformation of the group takes one point into another (active and passive views). Hence we have a group of automorphisms for the objective properties of points, or of geometry. Generalizing this idea to other geometries, any group of transformations may serve as the group of automorphisms of some geometry, hence the group defines or characterizes the geometry.

This same idea of relativity has been decisively used in all areas of physics since then. Physical processes happening around us are independent of observers (frames); all observers in a certain class are equally admissible, and these observers are related to each other by a group of transformations. Objective properties of the phenomena must be independent under this group of transformations which acts then as the group of automorphisms of the physical theory. Conversely, again, a transformation of the group makes one possible phenomena into another possible one, or we can generate new solutions from the old ones by symmetry transformations. Thus the group of automorphisms characterizes a physical theory or a physical law as it does a geometry.

If we are given a physical theory we can determine its symmetry by investigating the transformations of the underlying dynamical variables. Conversely, starting from the symmetry principles we can discover the form of the physical laws and their equations, as was indeed the case for special relativistic equations of mechanics, or thermodynamics, or the general relativistic equations of gravitation. The dynamical variables in physics are not only the coordinates of points as in geometry, but other entities like forces, velocities, fields, scales; and we must look among all the variables for the objective properties of phenomena. This involves all kind of transformations, beyond the coordinate transformations; for example, permutations of identical objects.

Table I lists, beginning with the euclidean geometry in $\mathbb{R}^3$, the geometry of some simple physical systems, the corresponding group of transformations (inertial observers), the objective properties (invariants), and physical laws. Note the euclidean geometry in $\mathbb{R}^3$ is the same as the physics of non-interacting nonrelativistic mass points *at rest*; the Galilean geometry that of moving particles, and the Minkowski geometry that of relativistic non-interacting mass points. In the last row we give the conformal transformations in the Minkowski space which we shall elaborate now in detail.

Scale and Conformal Transformations. Among homeometry transformations most closely related to geometrical transformations are scale transformations, not only an overall change of scale but also the change of scale from point to point in space-time. The transformations of the figures in the two-dimensional plane by a combination of relations, translations and constant dilatations is the simplest example.

$$x' = \rho(x \cos\theta + y \sin\theta + a)$$
$$y' = \rho(-x \sin\theta + y \cos\theta + b) \tag{1.1}$$

Simple examples of position-dependent scale transformations are given in Fig. 4 by the modifications of the periodic function sin x. Whereas y = A sin kt (or A sin kx) is the

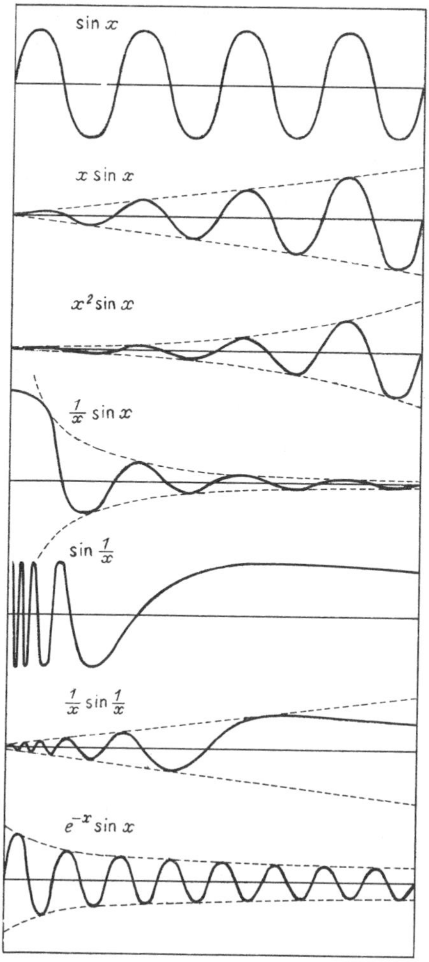

Fig. 4

solution of the equation of a simple periodic spring, $\ddot{y} = -k^2 y$, or of a one-dimensional stationary wave equation in continuum or quantum mechanics

$$y'' + k^2 y = 0 \tag{1.2}$$

the other functions in Fig. 4 occur as solution of a "dynamical" problem with "interactions":

$$y'' + k^2 y = F(y, y', x). \tag{1.3}$$

Thus, by a proper position-dependent scale transformation we can map the dynamical eq.(1.3) into the "kinematical" eq.(1.2). For example, an oscillator with damping

$$m\ddot{x} = -kx - R\dot{x}$$

has the solution

$$x(t) = Ce^{-t/\tau} \cos (\omega_1 t + \gamma)$$

where $\tau = 2m/R$ and $\omega_1^2 = \omega_o^2 - \tau^{-2}$; ω_1 is the natural frequency, while $\omega_o = (k/m)^{1/2}$ is the natural frequency of the free oscillator. Here the damping term provides the background for the agent which we can thus interprete as a natural 'scale' change. One can think of other similar environmental effects acting on the system.

In the general case, the group of space-time dependent scale transformations of a special kind in the Minkowski-space, the conformal group, is shown in the table, page 161. This interpretation of the special conformal transformations is by no means obvious

Physical or Geometrical System	Space of Events or State Space	Inertial or Equivalent Observers	Geometry. Invariants	Group of Auto-morphisms	Group Space	Intrinsic Parameters	Invariant equation of Motion
Space-point	$\mathbb{R}_3$	$\underset{\sim}{x}'=R\underset{\sim}{x}+\underset{\sim}{a}$	$(\underset{\sim}{x}_1-\underset{\sim}{x}_2)^2=$ =invar.	Euclidean group $\mathbb{E}_3$	6-dim. manifold $\mathbb{R}_3 \times S_4$		
non-relativistic mass-point	$\mathbb{R}_3 \times \mathbb{R}_1$	$\underset{\sim}{x}'=R\underset{\sim}{x}+\underset{\sim}{v}t+\underset{\sim}{a}$ $t'=t+t_o$	$t_2-t_1=$ =invar., $\lvert\vec{x}_1-\vec{x}_2\rvert=$ invar. if $t_2=t_1$	Galilei group $\mathbb{G}$	10-dim. manifold	Mass m	$m\ddot{\underset{\sim}{x}}(t)=\underset{\sim}{F}=\nabla\phi$ $\nabla^2\phi=\rho$
Space-time event (Relat. mass point)	$\mathbb{M}_4$	$x'^{\mu}=\Lambda^{\mu}_{\nu}x^{\nu}+a^{\mu}$	$(x_1-x_2)^2=$ =inv. or $ds^2=dx_{\mu}dx^{\mu}$	Poincaré group $\mathbb{P}$ and space-time reflections	10-dim. manifold	Rest mass m_o	$m_o\ddot{x}^{\mu}(s)=$ $=eF^{\mu\nu}\dot{x}_{\nu}$ $\Box A_{\mu}=j_{\mu}$
Space-time-scale point	$\tilde{\mathbb{M}}_4$	$x'^{\mu}=\Lambda^{\mu}_{\nu}x^{\nu}+a^{\mu}$ $x'^{\mu}=\rho x^{\mu}$ $x'^{\mu}=\dfrac{x^{\mu}+c^{\mu}x^2}{1+2c_{\mu}x^{\mu}+c^2x^2}$	Light cone= invar.	Conformal group $\mathbb{C}_4$	15-dim. manifold	m_o	
"	5-dim. cone in $\mathbb{R}_6$	$\eta'_A=L^B_A\,\eta_B$ A,B=1...6	$\eta_A\eta^A=$inv.	$O(4,2)\sim su(2,2)$	"	m_{oo}	$m_{oo}(\frac{du^A}{d\sigma}+\gamma^{A,BC}u_Bu_C)=k^A$
Space-time-gravitation	Riemannian space	$x'^{\mu}=x'^{\mu}(x)$	$ds^2=g_{\mu\nu}x^{\mu}x^{\nu}$			-	$\frac{du^{\mu}}{ds}=\Gamma^{\mu}_{\nu\sigma}u^{\nu}u^{\sigma}$ $G_{\mu\nu}=T_{\mu\nu}$

Table 1. The geometry of some simple physical systems, the corresponding group of transformations, the objective properties and the physical laws.

or universally accepted. Early interpretations of the conformal group were in terms of transformations into non-inertial accelerated frames. For our purposes we adopt the former point of view. Note also that the same abstract group may find different physical interpretations.

It follows from our definition of homeometry that if we enlarge the space of variables or the state space for our system, then homeometry transformations will look like geometric symmetry transformations again, but a generalized geometry in a generalized space. This is indeed the case for the conformal transformations if we go to the six-dimensional space (also shown in Table I). Let us introduce the overall scale K as the fifth coordinate, and the quantity $\lambda = K x^2$, where x^2 is the Minkowski-length as the sixth coordinate. If we define dimensionless coordinates $\eta^\mu = K x^\mu$, $\mu = 0,1,2,3$, and $\eta_4 \equiv -\frac{1}{2}(\frac{K}{\ell_o} + \ell_o \lambda)$, $\eta_6 \equiv -\frac{1}{2}(\frac{K}{\ell_o} - \ell_o \lambda)$, where ℓ_o is the same elementary length, then the non-linear conformal transformations become linear transformations in the six-dimensional pseudoeuclidean space.

Thus we may establish in the ordinary space equivalence between situations or events (symmetry) which otherwise do not look equivalent. The abstract larger space proves this mathematically. Usually we do not change scale from point to point neither in our experiments, nor theoretically. But nature apparently makes use of this in phenomena which we call dynamics.

For example, in a Kepler problem, certain orbits of the same energy are equivalent under rotations of the coordinate frame, a geometric transformation. Now orbits of *different* energies can also be made to be equivalent under special conformal scale-transformations (homeometry transformations).

In this generalized sense we extend all the significance and uses of symmtery to homeometry. Homeometry

(i) allows us to obtain from one solution another solution (differing in scales)

(ii) establishes equivalence between two situations (active) or between two observers (passive), (carrying different scales)

(iii) describes indistinguishability of two situations (of different scales)

(iv) expresses impossibility of measuring absolute coordinates *and scales*

(v) means invariance of physical laws under more general transformations

(vi) implies the existence of "constants of motions" (some of which may now be time-dependent!).

General Homeometry. It is now clear how to proceed in the general case. (a) We have to introduce all the variables or attributes of the system and define the state space, which may include but is larger than the 'geometric space'; (b) We have to identify the new transformations which map one situation into another.

Examples of homeometry abound in particular in biological world and in art. For some beautiful samples we refer to an article by K.L. Wolf [1949].

Quantum Theory. As is well known the implications of symmetry principles are more fundamental and decisive in quantum physics than in classical physics. Even new results and insight have been obtained via the guidance of the invariance principles. This is because the state space of quantum systems is a mathematical infinite-dimensional linear space of 'state vectors', and the group operations are represented by linear operators in this space, and we have the full power of the theory of group representations.

''States'' in classical physics are functions like $\vec{x}(t)$ or $\phi(\vec{x},t)$ representing trajectories or fields. In quantum theory the state vector ψ_t at time t is an element of a Hilbert space $\mathcal{H}$, e.g. $\mathcal{L}^2(\mathbf{R}^n)$, (or of a more general linear space). The state is a *ray* in $\mathcal{H}$, $\alpha\psi_t, |\alpha| = 1$. There exists a scalar product (ψ_t,ϕ_t) in $\mathcal{H}$, such that $|(\psi_t,\phi_t)|^2$, (independent of α), represent observable probabilities between the two states $\alpha\psi_t, \beta\phi_t$, which must be invariants of the theory, i.e. the same for all equivalent observers. The evolution of the system is governed by the equation $(\frac{i\partial\psi_t}{\partial t} = H\psi_t)$. These principles incorporate (i) the superposition principle of states, and (ii) conservation of total probability.

The symmetry transformations conserve the scalar products (ψ,ϕ) up to a phase, and are represented by unitary (or antiunitary) operators in $\mathcal{H}$, up to certain phase factors. Conversely, the state space $\mathcal{H}$ must be the carrier spaces of unitary representations of symmetry groups. The generators of symmetry transformations which are physical observables, are represented by self-adjoint operators in $\mathcal{H}$. Before we discuss the homeometry transformations, we review two examples from symmetry principles.

Example 1. For the Schrödinger equation for a free particle, $i\hbar \frac{\partial\psi}{\partial t} = -\frac{\hbar^2}{2m}\nabla^2\psi$, the space of all solutions carries a representation of the Galilei group (cf. Table I), more precisely of an extension of the Galilei group. The 10 differential operators $H = \frac{\partial}{\partial t}$, $\vec{P} = \vec{\nabla}$, $J = \vec{r} \times \vec{\nabla}$ and $\vec{K} = t\vec{\nabla}$ with commutation relations

$$[J_i, J_k] = \varepsilon_{ik\ell} J_\ell \; ; \quad [J_i, K_j] = \varepsilon_{ijk} K_k \, , \quad [J_i, P_j] = \varepsilon_{ijk} P_k \, ,$$

$$[K_i, H] = P_i \, , \quad [K_i, P_j] = \delta_{ij} \, m I \, ,$$

generators of the symmetry group, define the observables (energy, momentum, angular momentum) and constants of the motion: Note that the time-dependent quantities $\vec{K}$ are constants of the motion in the generalized sense:

$$\dot{K} = -\frac{\partial K}{\partial t} + [H, K] = 0,$$

whereas $[H,\vec{P}] = 0$, $[H,\vec{J}] = 0$, $[H,\vec{J}\cdot\vec{P}] = 0$ (but $[\vec{J}\cdot\vec{P},\vec{K}] = \vec{K}\times\vec{P}$). For a given particle P^2 and $(\vec{K}\times\vec{P})^2$ are Casimir operators;

their values are fixed and they characterize the representation and the state space. Under a Galilean transformation of coordinates shown in Table I, the state vector transforms as

$$\psi'(x,t) = e^{im(\vec{v}.R\vec{x}+\frac{1}{2}v^2t)+ic}\,\psi(x,t)$$

and the Schrödinger equation remains invariant.

Example 2. Many stationary problems in quantum theory reduce to an eigenvalue equation of the form $Hu = Eu$, e.g. $[-\frac{\hbar^2}{2m}\nabla^2 + V(\vec{r})]u(\vec{r}) = Eu(\vec{r})$. For an eigenvalue E, we find $\psi_t(E) = U_E e^{-\frac{i}{\hbar}Et}$. If there exist operators L_i such that $[H,L_i] = 0$, $[L_i,L_j] = c_{ij}^k L_k$, then L_i are generators of a *Lie-symmetry of the Hamiltonian H.* Then if u is a solution, $L_i u$ is also one; if at $t = 0$, L_i has a definite value $L_i\,\psi_o = m\psi_o$, then at time t, L_i has the same value, $L_i\psi_t = m\psi_t$. A Lie symmetry of H implies a degeneracy of the eigenvalues of H. Let $Hu_k^s = E_s u_k$, s = fixed, $k = 1...,d_s$. This d_s-dimensional eigenspace of E_s carries a representation T_g of a "degeneracy group" G, $[H,T_g] = 0$. The d_s-levels form *s-multiplets* of the same energy E_s. For example, (2j+1)-multiplets of atomic levels having the same angular momentum j and energy, or n^2-multiplets of the non-relativistic hydrogen atom of energy $E_n = -\frac{1}{2}\frac{\alpha^2 m}{n^2}$, $n = 1,2,3,...$.

The ideas in Examples 1 and 2 can be applied to relativistic problems, in particular to the Poincaré group (cf. Table I). The representations of the inhomogeneous Lorentz group play an important role in the classification of all relativistic particles, as well as in defining the kinematical frame (e.g. the state space) of all relativistic quantum theories.

Homeometry Groups in Quantum Theory. If we look at the spectrum of the H-atom

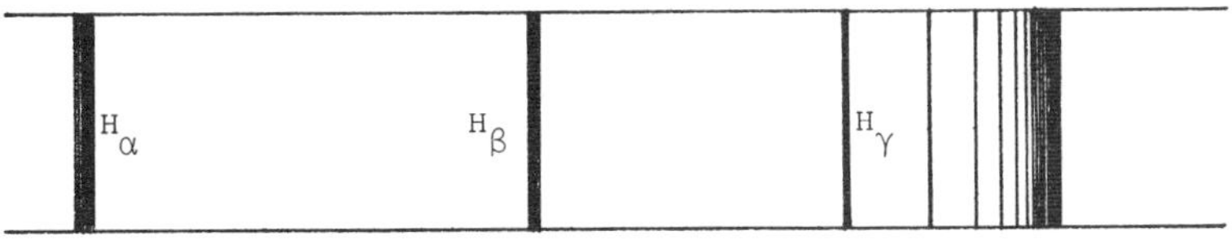

we recognize a homeometric pattern both in the spacing and intensity of lines, and not a symmetric pattern. Similarly the whole Periodic Table of Elements of D.I. Mendeleev and Lothar Meyer is a beautiful example of homeometry. The spectral lines connect different energy-levels. Thus, we must use transformations which connect different energy-levels in addition to those between degenerate levels. Consequently we go beyond the simple symmetry transformations of the Hamiltonian discussed in Example 2 above.

In the mathematical formulation of homeometry we just do not stop at the Lie algebra $\{L_i\}$ such that $[H,L_i] = 0$, but proceed as follows:

(i) Let $\vec{r}$, $\vec{p}$, ... be the fundamental dynamical variables of the theory (or more general objects), e.g. $\vec{r} = \vec{q}$, $\vec{p} = -i\hbar\vec{\nabla}$ $[p_i, q_j] = -i\hbar\,\delta_{ij}$.

(ii) Construct the enveloping field ε from these basic quantities, i.e. r^2, $\vec{r}\times\vec{p}$, p^2, r^{-1}, etc...

(iii) Let $\mathcal{L}$ be a Lie algebra in ε, and $\varepsilon_{\mathcal{L}}$ the enveloping algebra of $\mathcal{L}$, such that the Hamiltonian H is a function of $\mathcal{L}$ or $\varepsilon_{\mathcal{L}}$: $H = f(\varepsilon_{\mathcal{L}})$.

(iv) Then H acts in an irreducible carrier space of the Lie algebra $\mathcal{L}$. If we know the representation of $\mathcal{L}$, we also know the complete spectrum of H.

(v) If, further, the external interactions can also be expressed in $\varepsilon_{\mathcal{L}}$, their matrix elements can be evaluated in the same representation of $\mathcal{L}$, and $\mathcal{L}$ generates the dynamical or homeometry group of the system.

Example 3. Consider the Hamiltonian of the non-relativistic Coulomb problem: $H = \frac{1}{2m} p^2 - \frac{\alpha}{r}$. The operators

$$\Gamma_o = \frac{1}{2} (rp^2 + r)$$

$$\Gamma_4 = \frac{1}{2} (rp^2 - r) \tag{1.4}$$

$$T = \vec{r}.\vec{p} - i$$

Satisfy the commutation relations of the Lie algebra $0(2,1) \sim SU(1,1)$:

$$[\Gamma_o, \Gamma_4] = iT, \quad [\Gamma_4, T] = -i\Gamma_o, \quad [T, \Gamma_o] = i\Gamma_4 \tag{1.5}$$

with

$$Q^2 = \Gamma_o^2 - \Gamma_4^2 - T^2 = J^2 = (\vec{r}\times\vec{p})^2 \tag{1.6}$$

Hence with $\Theta = r(H - E)$ the Schrödinger equation can be written as

$$\Theta H\, \psi = [(\frac{1}{2m} - E)\Gamma_o + (\frac{1}{2m} + E)\Gamma_4 - \alpha]\psi = 0 \qquad (1.7)$$

Let $\tilde{\psi} \equiv e^{-i\theta T}$ with $\tanh\theta = \frac{E+m/2}{E-m/2}$. Then using (1.5) in (1.7) we get the transformed equation

$$(\sqrt{\frac{-2E}{m}}\,\Gamma_o - \alpha)\,\tilde{\psi} = 0 \qquad (1.8)$$

Now in the representation of (1.5) characterized by (1.6) [i.e. j(j+1)] the spectrum of Γ_o (which is always discrete) consists of the values

$$n = s + j + 1, \quad s = 0,1,2,\ldots$$

Hence $E < 0$ and

$$E_n = -\frac{\alpha^2 m}{2n^2} \,. \qquad (1.9)$$

Furthermore, in order to find degeneracy of levels, we have to add angular momentum $\vec{J} = (\vec{r}\times\vec{p})$ to the set (1.4), and the Lenz vector $\vec{A}$. If we do that we obtain a Lie algebra $\mathcal{L}$ with 15 elements: Γ_o, Γ_4, T, $\vec{J}$ and

$$\vec{A} = \frac{1}{2}\,\vec{r}p^2 - \vec{p}(\vec{r}.\vec{p}) - \frac{1}{2}\vec{r}$$

$$\vec{M} = \frac{1}{2}\,\vec{r}p^2 - \vec{p}(\vec{r}.\vec{p}) + \frac{1}{2}\vec{r} \qquad (1.10)$$

$$\vec{\Gamma} = r\vec{p}$$

which generate the conformal group* $0(4,2)$ [cf. Table I]. Note, however, that this conformal group acts in the four-dimensional momentum space, rather than the coordinate space-group shown in Table I.

(*Actually, a particular representation of $0(4,2)$ characterized by the value of the Casimir operators $Q_2 = -3$, $Q_3 = 0$, $Q_4 = -12$.)

Now we have obtained transformations which connect all levels of the H-atom. In order to see more clearly the homeometry character, consider again the equation

$$\left(-\frac{1}{2m}\nabla^2 - \frac{\alpha}{r} - E\right)\psi = 0 \tag{1.11}$$

One of our generators in eq. (1.4) and (1.10), namely

$$T = \vec{r}.\vec{p} - i$$

is a scale operator:

$$e^{-i\theta T}\, r\, e^{i\theta T} = r/r_o \; ; \qquad e^{-i\theta T}(r\nabla^2)e^{i\theta T} = r_o(r\nabla^2)$$

where $\theta = \ell n\ r_o$.

Thus, we can transform eq.(1.11) into

$$\left(-\frac{r_o^2}{2m}\nabla^2 - \frac{\alpha}{(r/r_o)} - E\right) e^{-i\theta T}\psi = 0$$

This equation compares different Kepler problems: Two problems with parameters (m,α) and $(\frac{m}{r_o^2}, r_o\alpha)$ have the same energy; or, if (m,α) has energy E, $(\frac{m}{r_o}, \alpha)$ has energy E/r_o, etc. If we are interested in mapping a level with quantum number n, $|n\rangle$, to another, $|n'\rangle$, we must use Lie algebra elements.

For example, $(\Gamma_4 \pm iT)$ increase the n-value of a state into $(n \pm 1)$. The group elements of the form $e^{i\varphi L}$, in general, map a level into a linear combination of all other levels. The same of course is true for symmetry: a rotation operator maps a level into a linear combination of all (2j+1)-levels. The main thing is that we have now identified the group of transformations between all levels. Of course, quantum mechanical states are different than the classical states. If we want to find the transformation mapping one "orbit" into another, we have to use coherent states, which are infinite superpositions of states $|n\rangle$. These come closest to the classical orbits.

Finally we note that the so-called "special functions" of mathematical physics which occur in the solutions of dynamical problems can all be related to representations of noncompact group of dynamical type (cf. Miller [1968], Vilenkin [1966], Tolman [1968]). This again indicates how such dynamical problems can be given a geometrical interpretation in agreement with our discussions.

Bibliographical and Historical Notes. The first application of group theory in natural sciences seems to go back to Hessel's classification of 32 crystallographic groups. About the same time E. Galois coined the word "group". The abstract group concept was introduced by Cayley (1854). About hundred years ago Sophus Lie began his investigations on the transformation groups (1873-1893), and the theory of Lie-groups is still alive and well today. Klein's Erlangen program (1873) and the idea of relativity has been elaborated in the text. Fourier analysis is seen today as a special case of harmonic analysis on groups. Frobenius (1896) and Schur (1903) have introduced the concept of representations of groups by linear transformations, so important today for quantum theory. With Lorentz and Poincaré (~1900) non-compact groups entered physics from the study of Maxwell's equations (Lorentz and Poincaré groups). Also the conformal group arose from Maxwell's equations(Bateman and Cunningham (1910). The implications of Riemannian geometry and symmetric spaces in physics were fully elaborated by Einstein and Cartan. Soon after quantum mechanics, the fundamental role of group representations in quantum theory was recognized by Weyl, Wigner and Van der Waerden. Finite and infinite dimensional representations of non-compact groups were used by Dirac (1928), Majorano (1932), Wigner (1931) and Bargmann (1947). Dynamical groups were introduced in 1964. Recent work involves applications of infinite-dimensional Lie algebras and groups and even locally non-compact groups. "It is no paradox that in our most theoretical moods we may be nearest to our most practical applications," (Alfred North Whitehead).

ACKNOWLEDGEMENTS

I should like to thank Prof. J. Brandmüller (Munich) whose lecture "Symmetry in Science and Art" inspired me to some of the thoughts in this paper.

Barut, A.O. [1972] *Dynamical Groups and Generalized Symmetries in Quantum Theory,* Christ Church (New Zealand): University of Canterbury Press.

Barut, A.O. and Rączka, R. [1977] *Theory of Group Representations and Applications,* Warsaw, Polish Scientific Publisher.

Englefield, M.J. [1972] *Group Theory and Coulomb Problem,* New York: Wiley and Sons.

Gilmore, R. [1974] *Lie Groups, Lie Algebras and some of their Applications,* New York: Wiley and Sons.

Hermann, R. [1966] *Lie Groups for Physicist,* New York: Benjamin.

Klein, F. [1893] *Math. Ann.*43, 63.

Loeble, E.M. (editor) [1968-75] *Group Theory and Applications,* Vol. I, II, III, New York: Academic Press.

Miller, W. [1968] *Lie Theory and Special Functions,*New York: Academic Press.

Miller, W. [1973] *Symmetry Groups and Applications*, New York: Academic Press.

Moshinsky, M. [1968] *Group Theory and Many-body Problems,*New York: Gordon and Breach.

Park, D. [1968] Resource Letter SP-1 on Symmetry in Physics, *Amer. J. Physics* 36, 1.

Tolman, J.D. [1968] *Special Functions: A Grouptheoretic Approach,* New York: Benjamin.

Vilenkin, N. Ja. [1966] *Special Functions and the Theory of Group Representations,* (Transl. of Amer. Math. Soc. Vol.22, Providence, R.I.).

Wolf, K.L. [1949] Symmetry and Polaritat, *Studium Generale* 2, 213.

Wybourne, B.G. [1974] *Classical Groups for Physicists,* New York: Wiley and Sons.

FIELD THEORIES WITH HIGHER DERIVATIVES AND CONFORMAL INVARIANCE*

H. P. Dürr
Max-Planck Institut für Physik und Astrophysik, München (Germany)

ABSTRACT

It is demonstrated that massless $(N+1)^{st}$-order Klein-Gordon equations for a scalar field and $(2N+1)^{st}$-order Weyl and Dirac equations for a spinor field are invariant under the 15-parameter conformal group if one attributes a (mass) dimension $(1-N)$ to the scalar field and a dimension $(3/2-N)$ to the spinor field. For $N > 0$ the canonical quantization of these field theories requires a vector space with indefinite metric. Only for $N=0$ in the scalar case and for $N=0$ and $N=1$ in the spinor case there exists a Hilbert subspace of normalizable eigenstates of the Hamiltonian which can be interpreted as physical state space.

The current operator is constructed in the spinor cases. For $N \geq 1$ conformal covariance of this current can only be achieved in a gauge invariant form of the theory where the gauge field is not an independent field but a bilinear form of the spinor fields.

Higher order derivative theories allow interactions which increase with distance and hence are of physical interest in connection with confinement and ultraviolet divergence problems.

* The lecture is essentially based on the paper by Chiang & Dürr [1975].

I. INTRODUCTION

Quantum field theories involving higher order derivatives have been studied quite early, in particular by Pais & Uhlenbeck [1950] and more recently e.g. by Barut & Mullen [1962]. Their physical interpretation, however, seemed to cause unsurmountable difficulties and hence they were not seriously pursued later on. Very recently theories of this type have again attracted considerable attention (see in particular Kauffmann [1974]; Kiskis [1974]) because of the possibility of exhibiting strong infrared singularities which may induce interactions *increasing* with distance. Such theories, hence, may be relevant for the 'small distance confinement' as discussed, at present by many authors (e.g. Johnson [1972]; Casher, Kogut & Susskind [1974]; Susslind & Kogut [1975]) in connection with the quark structure of hadrons. More important from our point of view, however, is their less singular behavior at small distances which allows interactions increasing more slowly as in conventional cases or even decreasing with decreasing distance. In the past Dürr [1974] and Bigi, Dürr & Winter [1974] showed that linear theories of this kind occur as canonical embeddings for fields with anomalous dimension smaller than the canonical dimension (they were called 'subcanonical' fields) which were used for the construction of non-divergent local theories with interaction, as e.g. in nonlinear spinor theories of the Heisenberg type which were initiated by Heisenberg [1953]. (For a review of the older work see e.g. Heisenberg & Dürr [1966]). For these considerations the masslessness and conformal invariance is only used as a feature emerging in the local limit where all masses appearing in the theory can be disregarded. Depending on the particular fashion in which these masses enter the infrared region the implications mentioned above may only by partially or completely irrelevant.

2. CONFORMAL INVARIANCE OF SCALAR FIELD EQUATIONS

We consider a massless scalar field which satisfies the $(N+1)^{st}$-order Klein-Gordon equation (N = integer)

$$\begin{aligned} K^{N+1}\varphi &= 0 \\ K &= -\partial^{\mu}\partial_{\mu} = -\Box \end{aligned} \qquad (2.1)$$

which can be formally derived from the Lagrangian density

$$\tilde{\mathcal{L}}^{(N)} = \varphi K^{N+1}\varphi \qquad (2.2)$$

To secure scale invariance of the action integral the field will be assumed to have the (mass) dimension, i.e. the intrinsic or scale dimension

$$d = \dim \varphi = 1 - N \qquad (2.3)$$

In this case one can prove now that the theory is also invariant under the 4-parameter special conformal transformations, and hence the full 15-parameter conformal group.

To demonstrate this it is convenient to rewrite the higher order derivative field equation (2.1) in canonical form involving only first order Klein-Gordon equations by introducing besides φ N additional independent fields $\underset{n}{\varphi}$ with $-N < n \leq N$ (n+N=even)

$$\left.\begin{aligned} K\varphi &= \underset{-N+2}{\varphi} \\ K\underset{-N+2}{\varphi} &= \underset{-N+4}{\varphi} \\ &\cdots\cdots \\ K\underset{N}{\varphi} &= 0 \end{aligned}\right\} \qquad (2.4)$$

or in the more compact form

$$\left.\begin{aligned} K\underset{n}{\varphi} &= \underset{n+2}{\varphi} \qquad -N\leq n\leq N;\quad n+N=\text{even} \\ \underset{-N}{\varphi} &= \varphi \qquad (\text{definition}) \\ \underset{N+2}{\varphi} &= 0 \end{aligned}\right\} \qquad (2.4')$$

The various fields φ_n have the dimension

$$d_n = \dim \varphi_n = 1 + n \tag{2.5}$$

By eliminating the N additional fields $n \neq -N$ one recovers the original equation. For N=2, e.g., one obtains from

$$\Box\Box\varphi = 0 \tag{2.6}$$

the set of equations $(\varphi \equiv \varphi_{-1})$

$$\left.\begin{aligned} -\Box\varphi_{-1} &= \varphi_{+1} \\ -\Box\varphi_{+1} &= 0 \end{aligned}\right\} \tag{2.7}$$

with

$$\begin{aligned} \dim \varphi_{-1} &= 0 \\ \dim \varphi_{+1} &= 2 \end{aligned} \tag{2.8}$$

The special conformal transformations are defined by the following nonlinear coordinate transformations

$$\begin{aligned} x^\mu \to x'^\mu &= \sigma^{-1}(c_1x)(x^\mu + c^\mu x^2) \\ \sigma(c_1x) &= 1 + 2cx + c^2x^2 \end{aligned} \tag{2.9}$$

involving the four Lie parameters c^μ of mass dimension one. The line element is not invariant under these transformations but varies as

$$ds^2 = dx^\mu dx_\mu \to ds'^2 = \sigma^{-2}(c_1x)ds^2 \tag{2.10}$$

exhibiting an x dependent dilatation. Only the light cone $ds^2 = 0$ is mapped onto itself. In some cases it is convenient to regard the special conformal transformation as a combination $\mathfrak{I} T \mathfrak{I}$ of an inversion $\mathfrak{I}$

$$x^{\mu} \overset{\mathfrak{I}}{\rightarrow} x'^{\mu} = \frac{x^{\mu}}{x^{2}} \tag{2.11}$$

i.e. a reflection on the unit 'circle', a subsequent translation T by c^{μ} and again an inversion. To demonstrate conformal invariance it hence suffices to show the invariance under inversion (2.11).

For our purpose we can restrict ourselves to the considerations of infinitesimal special conformal transformations involving infinitesimal Lie parameters c^{μ}. In this case the transformation law (2.9) can be simplified. In particular one finds for the coordinate differential

$$\begin{aligned} dx^{\mu} &\overset{c}{\rightarrow} dx'^{\mu} = dx^{\mu} + \delta dx^{\mu} \\ \delta dx^{\mu} &= -2(c\cdot x)dx^{\mu} + 2(c^{\mu}x_{\nu} - x^{\mu}c_{\nu})\, dx^{\nu} \end{aligned} \tag{2.12}$$

and the derivatives

$$\begin{aligned} \partial_{\mu} &\overset{c}{\rightarrow} \partial'_{\mu} = \partial_{\mu} + \delta\partial_{\mu} \\ \delta\partial_{\mu} &= 2(c\cdot x)\partial_{\mu} + 2(c_{\mu}x^{\nu} - x_{\mu}c^{\nu})\partial_{\nu} \end{aligned} \tag{2.13}$$

Under the infinitesimal special conformal transformation our fields φ_n, in general, transform inhomogeneously

$$\left.\begin{aligned} \varphi_n(x) &\overset{c}{\rightarrow} \varphi'_n(x') = \varphi_n(x) + \delta\varphi_n(x) \\ \delta\varphi_n(x) &= \bar{\delta}\varphi_n(x) + \Delta\varphi_n(x) \\ x^{\mu} &\overset{c}{\rightarrow} x'^{\mu} = x^{\mu} + \delta x^{\mu} \\ \delta x^{\mu} &= -2(c\cdot x)x^{\mu} + c^{\mu}x^{2} \end{aligned}\right\} \tag{2.14}$$

As is well-known (see e.g. Mack & Salam [1969]) Lorentz and scale invariance determine completely the homogeneous part of the transformation law of the field (finite dimensional representation of the little group), namely

$$\bar{\delta}\phi(x) = 2D(c\cdot x)\phi(x) + 2c_\lambda x_\rho S^{\lambda\rho}\phi(x) \tag{2.15}$$

where D = 'dimension' matrix with eigenvalues d and $S^{\lambda\rho}$ = spin tensor. Hence for the scalar field $\underset{n}{\varphi}$ of definite dimension d_n

$$\bar{\delta}\underset{n}{\varphi}(x) = 2d_n(c\cdot x)\underset{n}{\varphi} \tag{2.16}$$

For irreducible representations the inhomogeneous part $\Delta\varphi$ has to vanish. We now require

1) The original field $\varphi = \underset{-N}{\varphi}$ shall transform according to an irreducible representation, i.e.

$$\Delta\underset{-N}{\varphi} = 0 \tag{2.17}$$

2) The transformation law of the other fields $\underset{n}{\varphi}$ with $n \neq -N$ are defined in such a way as to guarantee for the form-invariance of the field equations (2.4).

On the basis of the second requirement one finds by induction a non-vanishing inhomogeneous part

$$\Delta\underset{n}{\varphi} = (N+n)(N-n+2)c\cdot\partial\underset{n-2}{\varphi} \tag{2.18}$$

To establish conformal invariance of the set of equations (2.4) it then suffices to show the conformal invariance of the *last* equation in (2.4), which requires

$$\Delta\underset{N+2}{\varphi} = 0 \tag{2.19}$$

This, Indeed, is verified from the general formula (2.18).

The transformation law of the (N+1) fields φ_n therefore has the structure

$$\delta \begin{bmatrix} \varphi_{-N} \\ \varphi_{-N+2} \\ \varphi_{-N+4} \\ \cdot \\ \cdot \\ \cdot \\ \varphi_{N-2} \\ \varphi_{N} \end{bmatrix} = c_\lambda [2Dx^\lambda + K^\lambda] \begin{bmatrix} \varphi_{-N} \\ \varphi_{-N+2} \\ \varphi_{-N+4} \\ \cdot \\ \cdot \\ \cdot \\ \varphi_{N-2} \\ \varphi_{N} \end{bmatrix} \qquad (2.20)$$

with

$$D = \begin{bmatrix} d_{-N} & & & & O \\ & d_{-N+2} & & & \\ & & \cdot & & \\ & & & \cdot & \\ & O & & & \cdot \\ & & & & & d_N \end{bmatrix} \qquad (2.21)$$

and

$$K^{\lambda} = \begin{bmatrix} 0 & & & & & \\ 4N & 0 & & & 0 & \\ & 8(N-1) & 0 & & & \\ & & \cdot & \cdot & & \\ & & & \cdot & \cdot & \\ & & & & \cdot & \cdot \\ 0 & & & & 4N & 0 \end{bmatrix} \partial^{\lambda} \tag{2.22}$$

Hence they transform according to a (N+1) dimensional *reducible but nondecomposable representation* of the local stability group of the conformal group which involve the nilpotent operator K^{λ} with the property $(K^{\lambda})^{N+1} = 0$. Since the algebra of the stability group has two nontrivial ideals such representations are field theoretically admissable as e.g. pointed out by Mack & Salam [1969].

3. QUANTIZATION AND STATE SPACE OF SCALAR THEORIES

The canonical form (2.4) of the field equations can be derived from the Lagrangian density (Chiang & Durr [1975])

$$\mathcal{L}^{(N)} = \frac{1}{2} \sum_{n=-N}^{N} [(\partial_{\mu}\varphi_{-n})(\partial^{\mu}\varphi_{n}) - \varphi_{-n}\varphi_{n+2}] \tag{3.1}$$

where the sum runs over $n+N$ = even between the limits and $\varphi_{N+2} = 0$ is implied. In this form quantization can be introduced in the conventional fashion. One finds

$$[\dot{\varphi}_{n}(x), \varphi_{m}(x')]_{t+t'} = -i\delta_{n,-m}\delta(\vec{x}-\vec{x}') \tag{3.2}$$

and all other equal time commutators being zero. From the antidiagonal form of these commutation rules one immediately deduces that for $N\neq 0$ the fields must be linear operators in a state space with *in*definite metric. For N = even there exists a single field, namely for n = 0, which has ordinary commutation rules, for N = odd there is none.

All the commutators can be deduced from the 2-point function of the original field $\varphi = \varphi_{-N}$ with lowest dimension

$$[\varphi_{-N}(x), \varphi_{-N}(x')] = \frac{i}{(2\pi)^4} \oint d^4p \frac{1}{(p^2)^{N+1}} e^{-ip\cdot(x-x')} \qquad (3.3)$$

which for $N \geq 1$ corresponds to

$$[\varphi_{-N}(x), \varphi_{-N}(x')] = -\frac{3i}{2\pi} \frac{(-)^N}{2^N} \frac{[(x-x')^2]^{N-1}}{(N-1)!(N+2)!} \varepsilon(x-x')\Theta((x-x')^2) \qquad (3.4)$$

The Feyman functions have a behaviour

$$<0|T\varphi_{-N}(x/2)\varphi_{-N}(-x/2)|0> \sim [x^2]^{N-1}\log[-x^2+i\varepsilon] \qquad (3.5)$$

which *increase* with distance and is related to the strong infrared singularity of the momentum integrals. The long distance features, of course, only show up in this way if the theory is not modified by mass terms.

The integrand in the momentum integrals

$$[p^2]^{-(N+1)} = [p^o-|\vec{p}|]^{-(N+1)}[p^o+|\vec{p}|]^{-(N+1)} \qquad (3.6)$$

indicates that there exists a (N+1)-pole for the positive and negative energy states. These energy 'multipoles' were first studied in the nonrelativistic case by K.L. Nagy [1966] as a generalization of the dipoles discussed by Heisenberg [1957]. They contain 'multipole ghost states' which are not eigenstates of energy and have an anomalous time behavior $t^m e^{-iEt}$.

For the construction of the quantum mechanical state space one has to introduce $N+1$ independent, momentum dependent annihilation and creation operators $a_n(\vec{p})$ and $a_n^*(\vec{p})$ $(-N \leq n \leq N,\ n+N = \text{even})$ which may be chosen such as to obey 'antidiagonal' commutation relations

$$[a_n(\vec{p}),\ a_m^*(\vec{p}\,')] = \delta_{n,-m}\,\delta(\vec{p}-\vec{p}\,') \tag{3.7}$$

and zero commutators for all other combinations. This means that the metric tensor in the one-'particle' subspace has only nonvanishing elements in the 'anti-diagonal'

$$\eta = \delta_{n,-m} = \left|\begin{matrix} & 0 & & & 1 \\ & & & 1 & \\ & & \cdot & & \\ & & \cdot & & \\ & \cdot & & & \\ & 1 & & 0 & \\ 1 & & & & \end{matrix}\right| \tag{3.8}$$

(disregarding momentum dependence). For N = odd the even $N+1$ different one-'particle' states therefore have *no* normalizable state but form $(N+1)/2$ *'null-ghost couples'* $(n,-n)$, i.e. pairs of non-orthogonal zero norm ghost states as discussed extensively by Dürr & Rudolph [1969,1970] and Dürr [1973]. For N = even there are besides $N/2$ of such 'null-ghost-couples' $(n,-n)(n\neq 0)$ normalizable states connected with the $n=0$ component.

The quantum mechanical state space can be constructed in conventional fashion by applying the various creation operators onto a vacuum state. All states $|N^{(-N)},\ldots, N^{(N)}>$ have zero norm except $|0,\ldots, N^{(0)},\ldots, 0>$ for N = even and states with equal occupation in an (n) and $(-n)$ states.

The field operators can be expanded in terms of these creation and destruction operators. For $N=1$ one has, e.g., the operators a_{-1}, a_{+1} and a^*_{-1}, a^*_{+1} and the expansions of

$\varphi = \varphi_{-1}$ and φ_{+1} are of the form

$$\left.\begin{aligned} \varphi_{-1}(x) &= \frac{1}{(2\pi)^{3/2}} \int \frac{d^3p}{(2p_o)^{1/2}} \frac{1}{2p_o} \left\{ [\, a_{-1} - a_{+1}(1+2ip_o t) e^{-ip\cdot x} + \text{'h.c.'} \right\} \\ \varphi_{+1}(x) &= \frac{1}{(2\pi)^{3/2}} \int \frac{d^3p}{(2p_o)^{1/2}} 2p_o \left\{ a_{+1} e^{-ip\cdot x} + \text{'h.c.'} \right\} \end{aligned}\right\} \quad (3.9)$$

In this case one deduces for the Hamiltonian

$$H^{(1)} = \int d^3p |\vec{p}| \, [\, a^*_{+1} a_{-1} + a^*_{-1} a_{+1} + 2\, a^*_{+1} a_{+1}] \quad (3.10)$$

with the property

$$\left.\begin{aligned} [H^{(1)}, a^*_{-1}] &= |\vec{p}| \, (a^*_{-1} + 2\, a^*_{+1}) \\ [H^{(1)}, a^*_{+1}] &= |\vec{p}| \, a^*_{+1} \end{aligned}\right\} \quad (3.11)$$

This shows that only the states $|0,N^{(+1)}>$ are eigenstates of energy. These states, however, have zero norm. Hence in this case there does not exist a physical Hilbert subspace spanned by the subset of δ-normalizable energy eigenstates.

One can show that also for arbitrary N only the 'highest' n-states generated by a^*_N are eigenstates of energy. Except for the conventional case N=0 these states, however, have vanishing norm. Hence for the massless higher order scalar theories, except for N=0, *there does not exist a physical Hilbert subspace* in the quantum mechanical state space.

This does *not* necessarily imply that theories of this type are physically irrelevant. They may be used as a limiting case for small distances for a physical theory involving masses.

The appearance of masses can drastically change the asymptotic state space. But even if the primary fields do not produce physical states an interaction of these fields may produce 'compounds' which relate to physical states. Such a theory hence would represent an example for a theory where the *constituents of physical states* do *not* show up asymptotically themselves, i.e. do not have a particle interpretation in the usual sense (quarks ?!).

The constituent field would behave, in fact, similar to the null-ghost fields in quantum electrodynamics in the description of Gupta [1950] and Bleuler [1950].

In quantum electrodynamics the vector potential $A_\mu(x)$ can be decomposed into the two components $A_\mu^{tr}(x)$ transverse to the direction of propagation k_μ for which there exists a particle interpretation, they are connected to photons, the longitudinal and time-like component $A_\ell(x)$ and $A_o(x)$, respectively, for which no *physical* particle interpretation exists. The vector field can be expanded in terms of annihilation and creation operators. The unphysical components generate unphysical longitudinal and time-like 'photons' from the vacuum state. The longitudinal 'photon' has positive norm, the time-like 'photon', however, negative norm. The metric tensor in the unphysical 1-particle subspace hence has the structure

$$\eta = \begin{bmatrix} 1 & 0 \\ 0 & -1 \end{bmatrix} \tag{3.12}$$

If one forms plus and minus superpositions of longitudinal and time-like 'photons' one obtains a 'null-ghost couple' consisting of two non-orthogonal norm zero states which is reflected in the structure of the metrical tensor

$$\eta = \begin{bmatrix} 0 & 1 \\ 1 & 0 \end{bmatrix} \tag{3.13}$$

These zero norm states, which we call 'good' and 'bad' ghosts, are annihilated by the destruction operators

$$\left.\begin{aligned} a_g &= \frac{1}{\sqrt{2}} [a_\ell + a_o] \\ a_b &= \frac{1}{\sqrt{2}} [a_\ell - a_o] \end{aligned}\right\} \qquad (3.14)$$

The subspace of physical states $|phys>$ is now defined by the Gupta condition

$$\left.\begin{aligned} \partial_\mu A^{\mu(+)}(x)\ |phys> &= 0 \\ \text{or} \qquad |\vec{k}|\, a_b(\vec{k})\ |phys> &= 0 \end{aligned}\right\} \qquad (3.15)$$

This indicates that the physical Hilbert space is simply given by the subspace which do not contain any bad ghosts. Since the good ghosts do not contribute (zero norm !) to any of these physical matrix elements they may be effectively omitted. The unphysical components of the vectors potential therefore have no projection in the physical subspace, they are not related to particle modes. Nonetheless their virtual presence gives rise to the 'Coulomb interaction' which therefore may be called a 'genuine interaction' or a 'non-particle-type interaction' if one reserves the concept of a 'particle' to asymptotically established modes.

As Dürr & Rudolph [1970] have shown a similar situation arises in linearized quantum gravitation theory. The ten components $k_{\mu\nu}(x)$ of the small deviation of the symmetric metric world tensor $g_{\mu\nu}(x)$ from the flat space Minkowski tensor can be decomposed into the two physical transverse-transverse components which relate to the two massless particle modes of spin 2 (gravitons) and eight unphysical components. The latter can be shown to group into four 'null-ghost-couples'. Their bad effects (violation of 'unitarity') is essentially avoided by requiring the 4-component Hilbert condition, the linearized form of the deDonder condition, for the physical subspace which is fulfilled if this space contains none of the (four) bad ghosts. The freedom of admixing the four 'good' ghosts is related to the general coordinate transformations. Because of the non-Abelian form of these transformations the 'unitarization' needs some additional sophistication (Faddeev-Popov ghosts) which I do not want to discuss here (Dürr & Rudolph [1972]).

4. CONFORMAL INVARIANCE OF SPINOR EQUATIONS

Conformal invariance of a third order Weyl spinor equation was first demonstrated by Dürr & van der Merwe [1974]. As a direct generalization of this case we consider a massless (2N+1)-order spinor equation for a 2-component Weyl spinor field

$$D(\bar{D}D)^N \psi = 0 \tag{4.1}$$

with the first order differential operators

$$\left.\begin{aligned} D &\equiv i\sigma^\mu \partial_\mu \\ \bar{D} &\equiv i\bar{\sigma}^\mu \partial_\mu \end{aligned}\right\} \text{ in short } D = i\partial\!\!\!/ \tag{4.2}$$

where $\sigma^\mu = (I, \vec{\sigma})$; $\bar{\sigma}^\mu = (I, -\vec{\sigma})$. The differential equations and their pseudohermitian conjugate may be formally derived from the Lagrangian density

$$\tilde{\mathcal{L}}^{(N)} = \frac{1}{2} \psi^* [D(\overrightarrow{\bar{D}D})^N + D(\overleftarrow{\bar{D}D})^N] \psi$$

For Weyl fields Lorentz invariance requires always an *odd* number of differential operators to occur in the Lagrangian. To establish scale invariance of the action integral the fields must have the (mass) dimension

$$d = \dim \psi = \dim \psi^* = \frac{3}{2} - N \tag{4.4}$$

Similar to the scalar case the higher order differential equation may be rewritten in canonical form involving only first order derivatives by introducing 2N additional subsidiary fields ψ_n $(-N < n \leq N)$ which obey the set of equation

$$\left.\begin{aligned} D\psi_n &= \psi_{n+1} \\ \psi_{-N} &= \psi \\ \psi_{N+1} &= 0 \end{aligned}\right\} \tag{4.5}$$

The fields have the dimension

$$d_n = \dim \psi_n = \dim \psi_n^* = \frac{3}{2} + n \tag{4.6}$$

and change their chirality at each step (N+n = even is right-handed, N+n = odd is left-handed).

If we require the original field $\psi = \psi_{-N}$ to transform according to an irreducible representation of the special conformal group the inhomogeneous term in the transformation law has to vanish

$$\Delta \psi_{-N} = 0 \tag{4.7}$$

Conformal invariance can be established if the condition holds

$$\Delta \psi_{N+1} = 0 \tag{4.8}$$

For the inhomogeneous term one deduces from the first equations in (4.5) for n+N = even:

$$\left.\begin{aligned} \Delta\psi_n &= (N+n)[(N-n+2)c\cdot\partial\,\psi_{n-2} + i\not{c}\,\psi_{n-1}] \\ \Delta\,\psi_{n+1} &= (N-n)[(N+n)c\cdot\partial\,\psi_{n-1} - i\not{c}\,\psi_n] \end{aligned}\right\} \tag{4.9}$$

From the second equation in (4.9) condition (4.8) can be immediately deduced which hence proves the conformal invariance of the spinor equations (4.1).

The variation of the spinor field has, therefore, the following structure

$$\delta \begin{bmatrix} \psi_{-N} \\ \psi_{-N+1} \\ \psi_{-N+2} \\ \vdots \\ \psi_N \end{bmatrix} = c_\lambda [T^\lambda + K^\lambda] \begin{bmatrix} \psi_{-N} \\ \psi_{-N+1} \\ \psi_{-N+2} \\ \vdots \\ \psi_N \end{bmatrix} \tag{4.10}$$

with the diagonal (reducible) part

$$T^{\lambda} = \begin{bmatrix} T^{\lambda}_{-N} & & & & & 0 \\ & T^{\lambda}_{-N+1} & & & & \\ & & T^{\lambda}_{-N+2} & & & \\ & & & \cdot & & \\ & & & & \cdot & \\ & & & & & \cdot \\ 0 & & & & & T^{\lambda}_{N} \end{bmatrix} \qquad (4.11)$$

$$T^{\lambda}_{n} = 2(\tfrac{3}{2} + n)x^{\lambda} + \sigma^{\lambda\rho}x_{\rho}$$

and the non-diagonal nil-potent part

$$K^{\lambda} = \begin{bmatrix} 0 & & & & & 0 \\ -2Ni\bar{\sigma}^{\lambda} & 0 & & & & \\ 4N\partial^{\lambda} & 2i\sigma^{\lambda} & 0 & & & \\ & 4(N-1)\partial^{\lambda} & 2(N-1)i\bar{\sigma}^{\lambda} & 0 & & \\ & \cdot & \cdot & \cdot & & \\ & & \cdot & \cdot & \cdot & \\ & & & \cdot & \cdot & \cdot \\ 0 & & & 4N\partial^{\lambda} & 2i\sigma^{\lambda} & 0 \end{bmatrix} \qquad (4.12)$$

with $(K^{\lambda})^{2N+1} = 0$.

The result can be immediately generalized to the (2N+1)-order *Dirac* spinor equations which can be interpreted as the combination of a right and a left chirality spinor field. Dirac Lagrangians, in contrast to Weyl Lagrangians may admit also an *even* number of derivatives which in the formalism above can be imitated by N = half-integer. In this case, however, conformal invariance can*not* be established, because the *first* equation in (4.9) has to be used for condition (4.8).

5. QUANTIZATION CONDITION AND STATE SPACE OF WEYL SPINOR THEORIES

Quantization and state space is best derived from the canonical form of the theory which is based on the Lagrangian ($\psi_{N+1} = 0$)

$$\mathcal{L}^{(N)} = \sum_{n=-N}^{N} \frac{1}{2} \{ \psi^*_{-n} \overleftarrow{D} \psi_n - \psi^*_{-n} \psi_{n+1} \} \tag{5.1}$$

from which the antidiagonal form of the anticommutator can be deduced

$$\{\psi_n(x), \psi^*_m(x')\}_{t=t'} = \delta_{n_1-m} \delta(\vec{x}-\vec{x}') \tag{5.2}$$

For the construction of the state space one has to introduce essentially 2(2N+1) independent, momentum dependent annihilation operators $a_n(\vec{p})$, $b_n(-\vec{p})$ referring to the annihilation of 'particles' and 'antiparticles', and the corresponding creation operators. For N+n = even the helicity of the 'particles' is positive, for N+n = odd it is negative. The operators can again be arranged to obey anticommutation rules which are simply 'antidiagonal'

$$\left.\begin{aligned} \left\{ a_n(\vec{p}), a^*_m(\vec{p}') \right\} &= \delta_{n_1-m} \delta(\vec{p}-\vec{p}') \\ \left\{ b_n(\vec{p}), b^*_m(\vec{p}') \right\} &= \delta_{n_1-m} \delta(\vec{p}-\vec{p}') \end{aligned}\right\} \tag{5.3}$$

and zero for all others, leading to an 'antidiagonal' metric tensor of the form (3.8) in the one-'particle' subspace. Since there are an odd number, namely (2N+1) different states ($N \leq n \leq N$, n = integer), there exists always one normalizable one-'particle' state with n = 0. The remaining 2N states form N 'null-ghost-couples'. The general state space can be constructed as usual from these states.

The Hamiltonian can be derived by common procedure. One derives that only the two 'highest' states with index n = N-1 and $n = N$ of opposite chirality are eigenstates of the Hamiltonian. As a consequence only the theory with $N = 0$ (usual massless Weyl Theory) and with $N = 1$ (3rd order theory) do contain a physical subspace. For all other theories the spinor fields have no projection on the physical subspace and hence do not lead to poles in the physical S-matrix.

The $N = 1$ theory with the three fields $\psi = \psi_{-1}$, $\hat{\psi} = \psi_{o}$ and $\hat{\hat{\psi}} = \psi_{+1}$ was studied extensively by Bigi, Dürr, Winter [1974] in the context of a canonical embedding of Heisenberg's nonlinear spinor theory (Heisenberg [1953, 1966]; Dürr et al. [1959]; Dürr [1961], [1966]). We will shortly come back to this in the next section.

The 2-point function of the original field $\psi = \psi_{-N}$ has the form

$$\langle 0|T\,\psi_{-N}(x/2)\,\psi^*_{-N}(-x/2)|0\rangle = \frac{i}{(2\pi)^4}\int d^4p\,\frac{\bar{\sigma}\cdot p}{(p^2)^{N+1}}\,e^{-ip\cdot x} \tag{5.4}$$

from which all other 2-point functions involving arbitrary field combinations $\psi_n\ \psi^*_m$ can be derived. In coordinate space the 2-point function (5.4) behaves for $N \geq 1$ as

$$\sim \bar{\sigma}\cdot\partial(x^2)^{N-1}\log(-x^2+i\varepsilon) = 2\bar{\sigma}\cdot x(x^2)^{N-2}[(N-1)\log(-x^2+i\varepsilon)+1] \tag{5.5}$$

i.e. for $N \geq 2$ vanishes for small distances and increases for large distances.

6. THE CURRENT IN WEYL SPINOR THEORIES AND ITS CONFORMAL TRANSFORMATION PROPERTY

The current for the $(2N+1)^{st}$ order Weyl theory is given by

$$j_\mu = \sum_{n=-N}^{N}{}' \psi^*_{-n}\sigma_\mu \psi_n + \sum_{n=-N}^{N-2}{}' \psi^*_{-n-1}\bar{\sigma}_\mu \psi_{n+1} \tag{6.1}$$

where the sum extends in even steps between the limits. Under infinitesimal special conformal transformations, on the basis of the variation of the spinor fields, this current is deduced to transform as

$$\begin{aligned} j_\mu(x) &\overset{c}{\rightarrow} j'_\mu(x') = j_\mu(x) + \delta j_\mu(x) \\ \delta j_\mu(x) &= \bar{\delta}\, j_\mu(x) + \Delta j_\mu(x) \\ x^\mu &\overset{c}{\rightarrow} x'^\mu = x^\mu + [-2(c\cdot x)x^\mu + c^\mu x^2] \end{aligned} \tag{6.2}$$

where the homogeneous part $\bar{\delta} j_\mu(x)$ of the variation is characteristic for a vector field of dimension 3

$$\bar{\delta} j_\mu(x) = 6(c\cdot x) j_\mu(x) + 2(c_\mu x^\nu - x_\mu c^\nu) j_\nu(x) \tag{6.3}$$

There does, however, exist, in general, an inhomogeneous part Δj_μ which may be cast into the simple form

$$\Delta j_\mu = c_\nu \partial^\nu V_\mu^{(N)} - c_\mu \partial^\nu V_\nu^{(N)} = 2\partial^\nu c_{[\nu} V_{\mu]}^{(N)} \tag{6.4}$$

involving a vector field of dimension 1 constructed from the spinor fields:

$$V_\mu^{(N)} = \sum_{n=-N}^{N} (N+n)[(N-n+2)\, \psi^*_{n-2}\sigma_\mu \psi_{-n} + (N-n)\, \psi^*_{n-1}\bar{\sigma}_\mu \psi_{-n-1}] \tag{6.5}$$

Due to the special form (6.4) of Δj_μ one immediately deduces

$$\partial^{\mu}\Delta j_{\mu} = 0 \tag{6.6}$$

which guarantees the conformal invariance of the continuity equation

$$\partial^{\mu}j_{\mu}(x) = 0 \tag{6.7}$$

Since Δj_o is a pure 3-divergence the time-independent charge operator

$$Q = \int d^3x j_o(x) \tag{6.8}$$

will also be conformal invariant.

Conformal covariance of the current operator is only required if the current itself occurs in the Lagrangian. This is the case in *gauge theories* where an interaction term of the form

$$\mathcal{L}_I = -gj_{\mu}A^{\mu} \tag{6.9}$$

is introduced containing a gauge field A^{μ}. The A^{μ} is assumed to transform irreducibly under conformal transformations as a vector field of dimension 1. As a consequence $\mathcal{L}_I$ can only be conformal invariant if j_{μ} transforms according to a *decomposable* representation of a field with dimension 3. This requires a vanishing of the inhomogeneous term Δj_{μ} which in turn implies

$$V_{\mu}^{(N)} = 0 \tag{6.10}$$

This condition seems to be fulfilled only for the conventional case $N=0$.

A more careful consideration of the field theoretical case (Dürr [1974]) where attention is paid to an appropriate definition of operator products at equal space-time points, reveals that this condition actually implies that, because j_{μ}

has to be constructed as the 'gauge invariant finite part' of the bilinear spinor operator products in (6.1), also the 'gauge invariant finite part' (designated by ${}^{\circ}_{\circ}\ {}^{\circ}_{\circ}$) of the bilinear spinor forms in (6.5) should be taken for the construction of the vector field. Hence condition (6.10) should be interpreted as

$$ {}^{\circ}_{\circ} V_{\mu}^{(N)} {}^{\circ}_{\circ} (x) = 0 \tag{6.11} $$

The 'gauge invariant finite part' of a spinor operator product can be obtained as a split-vector limit $\xi \to 0$ of the corresponding bilocal spinor field product involving the gauge field A_μ in the usual form $\exp[-ig\int_{x-\xi/2}^{x+\xi/2} A_\mu d\ell^\mu]$. For $N > 0$ condition (6.11) then states that the gauge field A_μ *cannot* be independent of the spinor fields. In fact, one derives from (6.11) the relation

$$ A_\mu(x) = -\frac{2\psi\pi^2}{N(N+1)(2N+1)g} \; {}^{\circ}_{\circ} V_{\mu}^{(N)} {}^{\circ}_{\circ} (x) \tag{6.12} $$

where ${}^{\circ}_{\circ}{}^{\circ}_{\circ}$ is now the 'ordinary finite part' prescription which for the terms in V_μ is given by

$$ {}^{\circ}_{\circ} \psi^{*}_{n-2} \sigma_{\mu} \psi_{-n} {}^{\circ}_{\circ} (x) = \overline{\lim_{\xi\to 0}} \left[T \psi^{*}_{n-2}(x-\xi/2)\sigma_{\mu} \psi_{-n}(x+\xi/2) - \frac{1}{4\pi^2 i} \frac{\xi_\mu}{\xi^2 i\varepsilon} \right] \tag{6.13} $$

with $\overline{\lim_{\xi\to 0}}$ the tetrad average of the split vector ξ and the subtraction term dictated by the propagator (5.4). By construction the $A_\mu(x)$ has the correct behaviour under gauge transformations

$$ A_\mu(x) \overset{G}{\to} A_\mu(x) + g^{-1}\partial_\mu \alpha(x) \tag{6.14} $$

where the inhomogeneous term arises as a direct consequence of the finite part prescription and the dimension 1 of V_μ.

By inserting (6.12) into (6.9) one obtains a nonlinear spinor expression in which the coupling constant g drops out. For convenience we may choose

$$g = 16\,\pi^2 \tag{6.15}$$

in which case (6.12) becomes

$$A_\mu(x) = -\frac{3}{2N(N+1)(2N+1)}\,\overset{\circ}{\circ}\,V_\mu^{(N)}\,\overset{\circ}{\circ}(x) \tag{6.16}$$

For N=1 we obtain

$$A_\mu(x) = -\frac{1}{4}\overset{\circ}{\circ}V_\mu^{(1)}\overset{\circ}{\circ}(x) = -\overset{\circ}{\circ}\psi^*_{-1}\sigma_\mu\psi_{-1}\overset{\circ}{\circ}(x) = -\overset{\circ}{\circ}\psi^*\sigma_\mu\psi\overset{\circ}{\circ}(x) \tag{6.17}$$

i.e. exactly the identification proposed earlier by Dürr & Winter [1972] which was the consequence of an attempt to incorporate gauge invariance (of the second kind) into a nonlinear spinor theory of the Heisenberg type *without* introducing an independent gauge field.

For N=2 one obtains similarly

$$A_\mu(x) = -\frac{1}{5}\overset{\circ}{\circ}[2\psi^*_{-2}\sigma_\mu\psi_o + \psi^*_{-1}\bar{\sigma}_\mu\psi_{-1} + 2\psi^*_o\sigma_\mu\psi_{-2}]\overset{\circ}{\circ}(x) \tag{6.18}$$

i.e. a vector potential of a structure similar to that discussed by Saller [1974]. In this case the gauge field has contributions from positive and negative chirality spinor components.

7. CONCLUDING REMARKS

In the latter section I have indicated a way how a nonlinear spinor theory involving *only a single Weyl spinor field* can be constructed which is *invariant under the full 15-parameter group and a gauge group* (Dürr [1974]). It is fully determined by these symmetries (no arbitrary coupling constants). Because of its conformal structure for small distances such a theory is formally renormalizable. The important problem left in such theories is to show the existence of a unitary S-matrix connecting in- and out-states which have to lie in a physical subspace of the state space. Since the ghost states enter the theory only in the form of 'null-ghost-couples' like in Gupta-Bleuler q.e.d., in the Yang-Mills-theory and the Heisenberg-Lee-model one may hope that this problem can be eventually solved. Quite a number of attempts have been made which look promising. For the unitarization of theories with dipole ghosts see in particular Heisenberg [1957], Karowski [1973] and Thalmeier [1974].

It is obvious that theories of this type may be generalized to incorporate higher internal symmetry groups, e.g. SU_2 or SU_3, and the corresponding gauge groups (Dürr & Winter [1972]). Hence they may very well offer a framework for the formulation of a fundamental theory of elementary particles.

REFERENCES

Barut, A.O. and Mullen, G.H. [1962] Ann. of Phys. 20, 203.

Bigi, I.I., Dürr, H.P. and Winter, N.J. [1974] Nuovo Cimento 22A, 420.

Bleuler, K. [1950] Helv. Phys. Acta. 23, 567.

Casher, A., Kogut. J. and Susskind, L. [1974] Phys. Rev. D10, 732.

Chiang, C.C. and Dürr, H.P. Preprint MPI-PAE/PTh 2 (Jan.1975), Max-Planck-Institut für Physik und Astrophysik, München.

Dürr, H.P., Heisenberg,W., Mitter, H., Schlieder, S., Yamazaki, K. [1959] Zeits. Naturforschg. 14a, 441.

Dürr, H.P. [1961] Zeits. Naturforschg 16a, 327.

Dürr, H.P. [1966] Acta Physica Austriaca, Suppl. III, 3.

Dürr, H.P. and Rudolph, E. [1969] Nuovo Cieento 62A, 411; [1970] 65A, 423; [1972] 10A, 597.

Dürr, H.P. and Winter, N.J. [1972] Nuovo Cimento 7A, 461.

Dürr, H.P. [1973] CPT-179, ORO-3992-130, Univ. of Texas Report.

Dürr, H.P. [1974] Nuovo Cimento 22A, 386.

Dürr, H.P. and P. du T. van der Merwe [1974] Nuovo Cimento 23A, 1.

Dürr, H.P., Preprint MPI-PAE/PTh 23/74; Max-Planck Institut fur Physik und Astrophysik, Münich.

Gupta, S.N. [1950] Proc. Phys. Soc. A63, 681.

Heisenberg, W. [1953] Nachr. Gottinger Akad. Wiss. IIa, 111.

Heisenberg, W. [1957] Nucl. Phys. 4, 532.

Heisenberg, W. [1966] Introduction to the Unified Field Thoery of Elementary Particles, Interscience Publ. London.

Johnson, K. [1972] Phys. Rev. D6, 1101.

Kauffmann, S.K., Quarks and Glue: Equation of motion; C.I.T. preprint 1974.

Karowski, M., Preprint: Dipole ghosts and unitarity (Jan. 1973), Max-Planck-Institut für Physik und Astrophysik, München.

Kiskis, J., Ph.D. Thesis Stanford University 1974.

Mack, G. and Salam, A. [1969] Ann. of Phys. 53, 174.

Nagy, K.L., State vector spaces with indefinite metric in quantum field theory, Akadémiai Kiakó, Budapest 1966.

Pais, A. and Uhlenbeck, G.E. [1950] Phys. Rev. 79, 145.

Saller, H. [1974] Nuovo Cimento 24A, 391.

Susskind, L. and Kogut, J., Preprint MPI-PAE/pTh 19/75, Max-Planck-Institut für Physik und Astrophysik, München.

Thalmeier, Diplom Thesis, Unitarization of dipole-regularized quantum field theories; University if Münich, 1974.

PHASE TRANSITIONS IN FIELD THEORIES

Asım Yıldız*

Lyman Laboratory of Physics, Harvard University
Cambridge, Mass.

ABSTRACT

Recent developments in phase transitions in field theories and symmetry restoration due to temperature, chemical potential and magnetic field are discussed.

1. INTRODUCTION

Unified field theoretical efforts of recent years have progressed from model building to field theoretical calculations. The successful renormalization of Yang-Mills fields generated much interest in the application of such field theoretical models.

Kirshnitz and Linde [1972] proposed the idea that the Weinberg model [1972] of unified electromagnetic and weak interactions may and can give field theoretical answers to some astrophysical problems. The specific question they raised was whether symmetries in field theories which are spontaneously broken can be restored either by increasing the temperature or

---

* Permanently at Mechanical Engineering Department of the University of New Hampshire, Durham, New Hampshire 03825

by applying large external fields. They observed the analogy between the energy expression of the Landau-Ginzburg [1950] theory

$$\varepsilon = \varepsilon_o + \frac{H^2}{8\pi} \frac{|(\nabla - 2ie\vec{A})\psi|^2}{2m} - \alpha|\psi|^2 + \beta|\psi|^4$$

and the Bose part of the Weinberg Lagrangian

$$\mathcal{L}_{Bose} = -\frac{1}{2}(\partial_\mu B_\nu - \partial_\nu B_\mu)^2 + |(\partial_\mu - 2igB_\mu)\phi|^2 + \mu^2\phi^2 - \lambda|\phi|^4$$

where ψ stands for the wave function of a Cooper pair, ε_o is the energy of a normal metal for vanishing magnetic field-H, m is the pair mass (whereas α and β are phenomenalogical parameters), B_μ is a vector field and ϕ is a complex scalar field. In each of these models, it is the opposing signs of the quadratic and quartic terms which give rise to the "condensate" i.e., to the non-vanishing vacuum expectation value. The qualitative estimations of Kirshnitz and Linde were quantitatively verified for a large temperature-T by Weinberg [1974] who considered both gauge and global symmetries for general renormalizable field theories. The restoration of the spontaneously broken symmetry in non-Abelian theories was shown to be possible using a non-zero chemical potential-μ for an absolutely conserved Fermion quantum number (cf. Harrington and Yildiz [1974]), and a large external magnetic field (cf. Salam and Strathdee [1974]). In each of these cases, the result was similar; when the disordering field (T,μ or H) is sufficiently strong (the strength is estimated by the coupling and masses of the scalar Bosons in the theory as well as by the electric charge in the external electromagnetic environment , a phase transition to a symmetric state is possible.

2. SPONTANEOUS SYMMETRY BREAKING

The concept that the physical states need not directly reflect the symmetries of an underlying dynamic has been one of the central themes in unified field theories. A spontaneously broken symmetry is a type of symmetry which can be obtained from an exactly symmetric Lagrangian provided that the physical vacuum is not invariant under the symmetry group. The development of spontaneously broken symmetries in particle physics was preceded by its application to problems in many body physics. A well-known example is the Heisenberg ferromagnet which is described as an infinite crystalline array of spin-1/2 magnetic dipoles, with spin-spin interactions with nearest neighbors such that neighboring dipoles tend to align. While the Hamiltonian of the system is rotationally invariant, the ground state is not. It is a state wherein all the dipoles are aligned randomly and which is infinitely degenerate for an infinite ferromagnet. Generalization of this quantum mechanical system to quantum field theories is obvious. The Hamiltonian will give its place to a Lagrangian and rotational invariance to an internal symmetry, and for the ground state of the ferromagnet, the vacuum state enters into the picture.

Thus, nature seems to possess certain symmetries such as isospin, strangeness and the Lagrangian possess these symmetries however not exactly. These type of symmetries can be obtained from the Lagrangian provided that the physical vacuum is not invariant under the symmetry group. That is to say, one can conjecture that the laws of nature can possess symmetries which are not obvious to the observer because the vacuum state is not invariant under them. This phenomena is described as "spontaneous breakdown of symmetry". It is instructive to begin by investigating spontaneous breakdown of symmetry in an example of a classical field theory. Thus, we turn our attention to O(2) scalar theory with quartic interaction which has the Lagrangian.

$$\mathcal{L} = \frac{1}{2}\,\partial_\mu\phi_i\partial^\mu\phi_i - \frac{1}{2}\,\mu^2\phi_i\phi_i - \lambda(\phi_i\phi_i)^2, \qquad i = 1,2 \tag{2.1}$$

where the parameters μ and λ stand for the bare mass and the bare coupling. We define the effective potential in the Lagrangian as

$$U(\phi_i) = \frac{1}{2}\,\mu^2\phi_i\phi_i + \lambda(\phi_i\phi_i)^2 \tag{2.2}$$

which has the symmetry of the theory and it is shown by Fig. 1. It is also obvious that we have a global gauge invariance which is implied by the transformation

$$\phi_i \rightarrow e^{ig\omega}\phi_i \tag{2.3}$$

where ω is a constant and g is the Hermitian charge matrix defined by

$$g = \begin{pmatrix} 0 & -i \\ i & 0 \end{pmatrix} \tag{2.4}$$

and we have chosen the field components ϕ_1 and ϕ_2 to be real. Expanding $\exp(ig\omega)$ in power series and noting the trigonometric identities, we may write the following form for eq.(2.3):

$$\begin{pmatrix} \phi_1 \\ \phi_2 \end{pmatrix} \rightarrow \begin{pmatrix} \cos\omega & \sin\omega \\ -\sin\omega & \cos\omega \end{pmatrix} \begin{pmatrix} \phi_1 \\ \phi_2 \end{pmatrix} \qquad (2.5)$$

This transformation is an invariant transformation of the field theory under consideration (global gauge transformation) and the group of such a continuous transformation is $O(2)$.

It is obvious that the potential $-U(\phi)$ does not have zeros except at the origin. If now the vacuum asymmetry conjecture is brought into consideration, the vacuum expectation value of the field-ϕ must have a non-zero value. Thus the effective potential must have minima other than at the origin. This can only be realized when the signs of the mass and coupling constant parameters become opposite. Thus, according to the description of the vacuum asymmetry conjecture we replace μ^2 by $-\mu^2$. This qualifies the effective potential-$U(\phi)$ to have minima at $\pm a$, $a = (1/2)(\mu/\sqrt{\lambda})$,

$$U(\phi) = \lambda(\phi^2 - a^2)^2 + \text{const.} \qquad (2.6)$$

where the constant term does not have any physical significance and can be disregarded (cf. Coleman [1973]).

The vacuum asymmetry has further consequences in the dynamics of the field. So far we have shown that, based on the vacuum asymmetry conjecture, the ground state of the system is degenerate at $\pm a$, and they are vacua. If we shift the origin to one of the minima

$$\phi' = \phi - a \qquad (2.7)$$

we require the vacuum expectation value-< > of the shifted field ϕ' to vanish at the minimum.

$$\langle\phi'\rangle = \langle\phi\rangle - a = 0 \tag{2.8}$$

and $\langle\phi\rangle = a$. Now, this is the vacuum and an observer does not notice the asymmetry at this minimum. The vacuum expectation value of the field vector can be equal to any of the degenerate values of the ground state. Since the vector-ϕ_i is a column matrix, then only one of the field components is required to have a non-zero vacuum expectation value. Furthermore O(2) invariance is still valid and the symmetry is a spontaneously broken symmetry.

After shifting the field, the potential reads

$$U = \lambda(\phi_1'^2 + \phi_2'^2 + 2a\phi_1')^2 \tag{2.9}$$

which implies that ϕ_1' has a mass-μ^2 as before. Furthermore ϕ_2' stays massless. Such a massless spin zero boson is called Goldstone boson. The name comes from the Goldstone theorem which reads: If a theory has a symmetry of the Lagrangian which is not a symmetry of the vacuum, then there exists a massless boson.

One knows that the Lagrangian-$\mathcal{L}(\phi_i, \partial_\mu\phi_i)$ of quantum electrodynamics of spinless electron which is an Abelian model obeys the gauge transformation

$$\phi_i \rightarrow e^{iq\omega}\phi_i \tag{2.10}$$

where q is the charge matrix defined by 2.4 and ϕ_i stands for real fields. The infinitesimal gauge transformation reads

$$\delta\phi_i = iq\phi_i\delta\omega . \qquad (2.11)$$

If $\delta\omega$ is considered to be space-time dependent, then the Abelian theory is not invariant under the prescribed transformation. Indeed

$$\delta(\partial_\mu\phi_i) = iq(\partial_\mu\phi_i)\delta\omega + iq\phi_i\partial_\mu(\delta\omega) \qquad (2.12)$$

where the last term violates the invariance. The remedy to revise the gauge invariance is to introduce a gauge field-A_μ by

$$\partial_\mu(\delta\omega) = -e\delta A_\mu \qquad (2.18)$$

where e is a free parameter (electric charge). The gauge field provides the generalization of the ordinary derivative to covariant derivative

$$D_\mu = \partial_\mu + ieqA_\mu \qquad (2.19)$$

in the Lagrangian $\mathcal{L} = \mathcal{L}(\phi_i, D_\mu\phi_i)$ which becomes gauge invariant since

$$\delta D_\mu\phi_i = iq(D_\mu\phi_i)\delta\omega. \qquad (2.15)$$

In this form of Lagrangian, A_μ does not appear as a dynamical variable. To make A_μ a dynamical variable we add the kinetic term $-1/4\ F_{\mu\nu}F^{\mu\nu}$ to the Lagrangian where $F_{\mu\nu} = \partial_\mu A_\nu - \partial_\nu A_\mu$ is the gauge field strength tensor.

Generalization to Yang-Mills is simply to provide a general internal symmetry group, a gauge group. Imitating the procedure of Abelian model we consider a theory that is invariant under the transformation

$$\delta\phi_i = T^a \, \delta\omega^a \phi_i \tag{2.16}$$

where $T^a_{(k\ell)}$ is the group parameter and $\delta\omega^a$ space-time dependent. Avoiding the repetition we can now write the covariant derivative

$$D_\mu \phi_i = (\partial_\mu + gT_a A^a_\mu)\phi_i \tag{2.17}$$

and g, similar to e, is a free parameter. The desired transformation properties of the gauge fields imply

$$\delta(D_\mu \phi_i) = T_a \, \delta\omega^a D_\mu \phi_i \tag{2.18}$$

with

$$\delta A^a_\mu = C^{abc} \delta\omega^b A^c_\mu - \frac{1}{g} \partial_\mu \delta\omega^a \tag{2.19}$$

and again the kinetic term $(-1/4 \, F^a_{\mu\nu} F^{a\mu\nu})$ need to be added to the Lagrangian $\mathcal{L}(\phi_i, D_\mu \phi_i)$. In Abelian case

$$[D_\mu, D_\nu] = ieqF_{\mu\nu} \tag{2.20}$$

and the gauge invariance of $F_{\mu\nu}$ is manifest. In non-Abelian (Yang-Mills) situation, we have

$$[D_\mu, D_\nu] = T_a F^a_{\mu\nu} \tag{2.21}$$

where

$$F^a_{\mu\nu} = \partial_\mu A^a_\nu - \partial_\nu A^a_\mu + gC^{abc} A^b_\mu A^c_\nu. \qquad (2.22)$$

Here we observe that addition to A_μ and ϕ_i interaction (electromagnetic case), we have A_μ, A_ν interaction.

3. RESTORATIONS OF SYMMETRIES (PHASE TRANSITIONS)

We shall now discuss when symmetries (of field theories) are broken spontaneously, these broken symmetries can be restored by increasing the temperature or by applying large external fields or by introducing a chemical potential. Restoration of symmetries require the vanishing effective mass which implies a phase transition in the field theory under consideration.

We start with temperature effects first. We have already discussed that due to spontaneous symmetry breakdown in ϕ^4-theory, effective potential develops two minima (see Fig. 2, page 229) at zero temperature T = 0.

Imbedding the system in a temperature bath we find the effective mass to become a function of temperature and thus for a critical value of temperature (critical temperature-T_c) the effective mass vanishes establishing the restoration of symmetry in the field (see Fig. 3, page 229) at critical temperature and the above.

Inspired by the non-relativistic many-body formalism we introduce the temperature by going to Euclidean domain and by the fourth component (time component) introduce the temperature parameter into the formalism. Technical details of such a formalism, definitions of Green's functions and other pertinent knowledge are given in technical papers (cf. Weinberg [1974], Dolan and Jackiw [1974], Harrington and Yildiz [1974], Bernard [1974]). In order to demonstrate the restoration of symmetry we thus are required to calculate the quantum effects in effective mass or in effective potential expressions. Classical terms (tree diagrams) are responsible for the generation of mass(es) and quantum fluctuations (loop diagrams) add up negatively to vanish the effective mass at a critical temperature. Thus, the temperature accelerates the negative contributions, namely one-loop contributions from the quantum fluctuations and only lowest order quantum contributions namely one-loop contributions, achieve the restoration of the symmetry or the phase transition. Second and higher loops become negligibly small. Since we are working within a framework of a weak coupling theory, characterized by a small dimensionless coupling constant e, then symmetries which are broken in lowest order should remain broken in higher order unless the perturbation scheme fails. This precisely happens at finite temperature when powers of $e\theta$ become of the same magnitude as the boson masses in the theory. To see this, we consider a loop expansion, where we have a factor of e^{2L} in a graph with L loops. To determine the powers of θ contributed by each loop, we examine a single loop with superficial degree of divergence D. Rescaling all dimensional factors in the single loop integration by θ ($\theta = -/kT$, k stands for the Boltzmann constant), we have the contribution

$$\theta^{D}\, I(P_{ext}/\theta,\ W_{ext}/\theta,\ m_{int}/\theta) \qquad (3.1)$$

where P_{ext} and W_{ext} represent the set of external momenta and energies, and m_{int} represents the internal masses. When $D < 1$, there are infrared divergences but the one loop contribution can grow no faster than θ as $\theta \to \infty$. When $D \geq 1$, there are no infrared divergences so that the loop goes θ^D for $\theta \to \infty$. Thus, in renormalizable theories in which the maximum value of D is 2 except for the quartically divergent contribution to the effective potential itself, we consider only graphs which are quadratically divergent. The leading finite temperature contribution of such graphs would be of the order $e^2\theta^2$. But all such quadratic divergences and $e^2\theta^2$ terms can be absorbed into a redefinition (or renormalization) of the boson mass. Thus, we will have an effective, temperature dependent boson mass (or inverse correlation length) which, if it does vanish, will vanish when $\theta_{crit} \sim m/e$, where m is the zero-temperature boson mass. In typical gauge theories of the weak and electromagnetic interactions, m/e is of the order of $G_F^{-i/2}$ (G_F - Fermi coupling) so that $\theta_{crit} \sim 300$ GeV $\sim 10^{15}$°K.

To be more specific we consider the most general, four-dimensional renormalizable Lagrangian which possesses a gauge invariance with respect to some compact semisimple Lie group G. The Lagrangian contains a multiplet of real spin zero fields $\phi_i(x)$ transforming according to a representation D_B of G, a multiplet of spin-1/2 fields $\psi_n(x)$ transforming according to a representation D_F of G and a set of spin-one gauge fields A_μ^α transforming according to the adjoint representation of G. The usual dimensional or power counting arguments restrict the Lagrangian to be of the form:

$$\mathcal{L} = -\frac{1}{4} F^a_{\mu\nu} F^{a\mu\nu} - \frac{1}{2}(D_\mu\phi)_i (D^\mu\phi)_i - \bar{\psi}\gamma^\mu D_\mu\psi - \bar{\psi} m_o \psi - P(\phi) - \bar{\psi}\Gamma_i \psi\phi_i \qquad (3.2)$$

where the gauge-covariant curl of the vector field A^a_μ is given by eq. 2.22, the gauge covariant derivative of the scalar field ϕ_i and the gauge-covariant derivative of the spinor field ψ_n are respectively described by

$$(D_\mu\phi)_i = (\partial_\mu\delta_{ij} + i(\theta_a)_{ij}A^a_\mu)\phi_j \qquad (3.3)$$

$$(D_\mu\phi)_i = (\partial_\mu\delta_{ij} + i(t_a)_{ij}A^a_\mu)\psi_i$$

where $(\theta_a)_{ij}$ is the matrix representing the a'th generator of G in the representation D_B. The matrix θ_a is proportional to the gauge coupling constants, and satisfies the antisymmetry, Hermiticity and commutation relations:

$$(\theta_a)_{ij} = (\theta_a)^*_{ij} = -(\theta_a)_{ji} \qquad (3.4a)$$

$$[\theta_a, \theta_b] = i\, C_{abc}\theta_c \qquad (3.4b)$$

where C_{abc} are a set of real totally antisymmetric structure functions, proportional to one or more gauge coupling constants. The matrix $(t_a)_{ij}$ is likewise proportional to the various gauge coupling constants and represent the a'th generator of G in the representation D_F. The matrix t_a may include terms proportional to γ_5 as well as 1 and satisfies the Hermiticity and commutation relations. Furthermore, the Yukawa coupling constants are included in the Yukawa matrices Γ_i which may also contain terms proportional to γ_5.

$P(\phi)$ is a real, polynomial function of ϕ, restricted to no greater than the fourth power in ϕ by renormalizability. Gauge invariance imposes the constraint

$$\frac{\partial P(\phi)}{\partial \phi_i} (\theta_a)_{ij} \phi_j = 0 \tag{3.5}$$

The G invariance of the Lagrangian is spontaneously broken by asserting that at least some of the ϕ_i have a non-vanishing expectation value, determined in lowest order by

$$\left. \frac{\partial P(\phi)}{\partial \phi_i} \right|_{\phi=\lambda} = 0 \tag{3.6}$$

where λ is the zeroth order vacuum expectation value. The perturbation expansion is then defined by using a shifted field $\phi_i'(= \phi_i - \lambda_i)$ which has a vanishing vacuum expectation value. The spin zero masses to zeroth order (tree approximation) read

$$M_{ij}^2 = \left. \frac{\partial^2 P(\phi)}{\partial \phi_i \partial \phi_j} \right|_{\phi=\lambda} \tag{3.7}$$

while the spin-1/2 masses become

$$m = m_o + \Gamma_i \lambda_i \tag{3.8}$$

and the spin-1 masses will be written as

$$\mu_{ab}^2 = -(\theta_a \lambda)_i (\theta_b \lambda)_i \tag{3.9}$$

The shifted Lagrangian will contain quadratic ($\mathcal{L}_{quad}$) and interaction ($\mathcal{L}'$) parts. Since $\mathcal{L}'$ contains derivative interactions, the quantization in the ξ gauge requires the introduction of a complex, spin-0 fermion ghost field ω_a. Disregarding the technical details we prefer to focus quickly to the determination of the leading temperature effects by isolating the quadratically divergent contribution to the effective boson mass. The one-loop contribution to the effective potential at zero temperature is given by

$$V_1 = -\frac{i}{2}\int\frac{(dk)}{(2\pi)^4}\ \mathrm{tr}\ \ell n(k^2+M^2) - 3\,\frac{i}{2}\int\frac{(dk)}{(2\pi)^4}\ \mathrm{tr}\ \ell n(k^2+\mu^2)$$

$$+\ 4\,\frac{i}{2}\int\frac{(dk)}{(2\pi)^4}\ \mathrm{tr}\ \ell n(k^2+m^2) \qquad (3.10)$$

The coefficient of 3 for the vector meson term relative to the scalar meson term is a reflection of its available polarization states while the factor 4 in the fermion term indicates that the trace over the Dirac space has been performed. Since eq. (3.10) stands quartically divergent, renormalization counter-terms are needed. These will be determined by insisting that the one-loop expansion is a valid perturbative expansion. Since it is simpler to impose this condition on the curvature of the effective potential and isolate the quadratically divergent contribution

$$\left.\frac{\partial V_1}{\partial\lambda_i}\right|_\infty = [-\frac{1}{2}f_{ikk}+\mathrm{tr}(\Gamma_i\gamma_4\Gamma_j\gamma_4)\lambda_j-3(\theta_\alpha\theta_\alpha\lambda)_i](i)\int\frac{(dk)}{(2\pi)^4}\frac{1}{k^2} \qquad (3.11)$$

where

$$f_{ikk} = f_{ijkk}\,\lambda_j \;, \qquad (3.12a)$$

$$f_{ijkl} = \frac{\partial^4 P(\phi)}{\partial\phi_i \partial\phi_j \partial\phi_k \partial\phi_\ell} \qquad (3.12b)$$

The quadratically divergent contribution to the effective boson mass is then

$$\frac{\partial^2 V_1}{\partial\lambda_i \partial\lambda_j} = [-\tfrac{1}{2} f_{ijkk} + tr(\Gamma_i\gamma_4\Gamma_i\gamma_4) - 3(\theta_\alpha\theta_\alpha)_{ij}]\,(i) \int \frac{(dk)}{(2\pi)^4}\,\frac{1}{k^2} \qquad (3.13)$$

This divergence is absorbed into an effective boson mass squared $M^2_{ij}(\theta)$:

$$M^2_{ij}(\theta) = M^2_{ij} + [\tfrac{1}{2} f_{ijkk} - tr(\Gamma_i\gamma_4\Gamma_i\gamma_4) + 3(\theta_a\theta_a)_{ij}](-i) \int \frac{(dk)}{(2\pi)^4}\,\frac{1}{k^2} \qquad (3.14)$$

The finite temperature evaluation of the quadratically divergent integral is accomplished with substitutions rules

$$\int \frac{(dk)}{(2\pi)^4}\,\frac{1}{k^2} = i\,\theta \sum_{n=-\infty}^{+\infty} \int \frac{(d\vec{k})}{(2\pi)^3}\,\frac{1}{\vec{k}^2 + (\omega_n \pi\theta)^2} \qquad (3.15)$$

where ω_n is $(2n+1)$ for bosons and $2n$ for fermions. Finally, the *effective boson mass* squared is

$$M^2_{ij}(\theta) = (M^2_R)_{ij} + \frac{\theta^2}{24}\,[f_{ijkk} + tr(\Gamma_i\gamma_4\Gamma_i\gamma_4) + 6(\theta_a\theta_a)_{ij}] \qquad (3.16)$$

where $(M_R^2)_{ij}$ is the zero temperature renormalized boson mass. We note that since $f_{ijk\ell}$ is of order e^2 while Γ_i and θ_a are of order e, then the temperature dependent contribution is of order $e^2\theta^2$, as expected. Furthermore, because of the Hermiticity conditions, the vector and spinor contributions to $M_{ij}^2(\theta)$ are positive definite while the scalar term f_{ijkk} need not be positive definite. Thus, in the case of spontaneous symmetry breaking when at least one of the eigenvalues of $(M_R^2)_{ij}$ is negative, the spinor and vector contributions at finite temperature will tend to restore the symmetry while the scalars may or may not aid in inducing symmetry restoration, depending on the sign of f_{ijkk}. Also it is possible that the scalar contribution can break a symmetry as the temperature is raised.

We now consider the Lagrangian (eq. 2.2) at finite temperature and incorporate a fermion chemical potential (cf. Harrington and Yildiz [1974]) as a Lagrange multiplier of the number density η given by

$$\eta = \psi^+\psi = \bar{\psi}\gamma_4\psi \tag{3.17}$$

Thus the chemical potential-μ will alter the fourth component of the fermion propagator, which, at finite temperature, implies the replacement of the discrete energy $\varepsilon_n = (2n+1)\pi\theta$, n an integer, by $\varepsilon_n - i\mu$. It is interesting to discuss the limit in which the fermion chemical potential μ is much larger than the temperature, i.e., $\bar{\mu} = \mu/\theta >> 1$, or the limit in which powers of μ compensate powers of e. To isolate the leading terms to any graph for e small and μ large, and $\bar{\mu} >> 1$, we again consider a loop expansion. The powers of μ contributed by each loop can be found by concentrating on a single loop with superficial degree of divergence D. We rescale by θ all dimensional factors in the single-loop integration so that the whole loop takes the form

$$\theta^D I(P_{ext}/\theta, \omega_{ext}/\theta, m_{int}/\theta, \bar{\mu}) \tag{3.18}$$

Now the single-loop integration will contain a Fermi distribution function in the integrand of the form

$$F(\frac{\varepsilon_p}{\theta}, \bar{\mu}) = \frac{1}{e^{(\varepsilon_p - \mu)/\theta} + 1} + \frac{1}{e^{(\varepsilon_p + \mu)/\theta} + 1} \tag{3.19}$$

where $\varepsilon_p = (p^2 + m^2)^{1/2}$ and m is the eigenvalue of the fermion mass matrix. Since $\bar{\mu} >> 1$, the second term in eq.(3.19) will act as an exponentially damping factor in the loop momentum p, while the first term will begin providing an exponential falloff in p when $\varepsilon >> \mu$, i.e., above the Fermi energy. We are primarily interested in μ large (~300 GeV) and thus much greater than m, so that the integral in 3.17 is effectively cut off when $p \sim \mu$. Thus expression 3.18 reduces to

$$\theta^D \bar{\mu}^D = \mu^D . \tag{3.20}$$

An effective, chemical potential dependent, boson mass can then be defined to absorb the leading $e^2\mu^2$ terms, and we anticipate a phase transition when $\mu_{crit} \sim m/e$. Again, in typical gauge theories of the weak and electromagnetic interactions, this transition would occur for $\mu_{crit} \approx 300$ GeV. Dimensionally, for $\bar{\mu} >> 1$, we expect $\mu \sim n^{1/3}$, where n is the fermion number density, so that the critical fermion number density would be of the order $n_{crit} \sim 10^{48}/cm^3$.

The calculation proceeds; by redefining the quadratic terms in the potential $P(\phi)$, we can absorb all quadratic divergences and the leading chemical potential effects

$$P_{eff}(\phi) = P(\phi) + \frac{1}{2} Q_{ij}(\mu)\phi_i\phi_j \qquad (3.21)$$

and by adding a compensating counter term to the Lagrangian

$$\delta \mathcal{L}' = \frac{1}{2} Q_{ij}(\mu)\phi_i\phi_j \qquad (3.22)$$

Here, $Q_{ij}(\mu)$ is some gauge invariant quadratically divergent matrix which also absorbs the expected $e^2\mu^2$ terms. Since the chemical potential enters the calculation only through the fermion propagators, the contributions of the fermion-loop tadpoles and the fermion-loop contributions to the boson self-energy will be considered.

The fermion-loop tadpole at finite temperature and chemical potential is given by

$$T_i^{\psi} = i(2\pi)^4 \; tr\{\Gamma_i\gamma_4\Gamma_j\gamma_4\} \; \lambda_j \; I_F(\theta,\mu) \qquad (3.23)$$

where

$$I_F(\theta,\mu) = \theta \sum_{n=-\infty}^{+\infty} \int \frac{d^3p}{(2\pi)^3} \frac{1}{\varepsilon_p^2+(\varepsilon_n-i\mu)^2} \tag{3.24}$$

The summation (over n) is evaluated with the use

$$\sum_{n=-\infty}^{+\infty} \frac{1}{\varepsilon_k^2+(\varepsilon_n-i\mu)^2} = \frac{1}{4\theta\varepsilon_k} [\tanh \frac{1}{2\theta} (\varepsilon_k+\mu)+\tanh \frac{1}{2\theta} (\varepsilon_k-\mu)] \tag{3.25}$$

and the zero temperature, zero chemical potential terms are separated

$$I_F(\theta,\mu) = I_F(0,0) + I_F^R(\theta,\mu) \tag{3.26}$$

where the divergent integral $I_F(0,0)$ is given by

$$I_F(0,0) = \frac{1}{2} \int \frac{(d\vec{k})}{(2\pi)^3} \frac{1}{\varepsilon_k} \tag{3.27}$$

while the temperature and chemical potential dependent term is finite

$$I_F^R(\theta,\mu) \cong -\frac{\theta^2}{4\pi^2} \int_o^\infty dx\; x\left(\frac{1}{e^{x+\bar{\mu}}+1} + \frac{1}{e^{x-\bar{\mu}}+1}\right) \tag{3.28}$$

The preceding integral can be done exactly

$$I_F^R(\theta,\mu) = -(\frac{\theta^2}{24} + \frac{\mu^2}{8\pi^2}) \tag{3.29}$$

But the counter term $\delta\mathcal{L}'$ contributes a term

$$\delta T_i^\psi = i(2\pi)^4 Q_{ij}(\mu)\lambda_j \tag{3.30}$$

so that we have for the finite part of $Q_{ij}(\mu)$ in the limit $\bar{\mu} >> 1$

$$Q^{fin}_{ij}(\mu) = (\mu^2/8\pi^2)\ \mathrm{tr}\ \{\Gamma_i\gamma_4\Gamma_j\gamma_4\}. \qquad (3.31)$$

Thus, the effective scalar boson mass matrix is given by

$$[m^2(\mu)]_{ij} = [m^2_R]_{ij} + \frac{\mu^2}{8\pi^2}\ \mathrm{tr}\ \{\Gamma_i\gamma_4\Gamma_j\gamma_4\} \qquad (3.32)$$

where the chemical potential provides a positive definite contribution. The vanishing of $m^2(\mu)$ determines μ_{crit} in the one-loop approximation. In the case of a spontaneously broken symmetry where $m^2_R < 0$, we have $\mu_{crit} \sim m_R/e$ since the Yukawa coupling matrices are each proportional to e.

We can equivalently express our result in terms of a critical number density n_{crit}, where n is the number of fermions minus the number of antifermions in a unit volume V:

$$n = (1/v) \int d^3x\ \eta(\vec{x}) \qquad (3.33)$$

and we write the renormalized (at $\theta = \mu = 0$) number density

$$n_R(\theta,\mu) \cong \frac{\theta^3}{\pi^2} \int_o^\infty dx\ x^2\left(\frac{1}{e^{x-\bar{\mu}}+1} + \frac{1}{e^{x+\bar{\mu}}+1}\right) \qquad (3.34)$$

where the fermion mass is again negligible compared to μ. This integral can be done exactly

$$n_R(\theta,\mu) \cong \frac{\theta^3}{3\pi^2}\ (\bar{\mu}^3 + \pi^2\bar{\mu}) \qquad (3.35)$$

Inverting this last expression, we write

$$\mu \simeq (3\pi^2 n_R)^{1/3}[1+O(\frac{1}{\mu})]$$

thus confirming the dimensional argument $(\mu \sim n^{1/3})$ leading to the estimate of the critical fermion number of $10^{48}/cm^3$.

Symmetry restorations by external magnetic fields for general physically interesting, renormalizable, four-dimensional models have not met much success (cf. Salam and Strathdee [1974]). Only positive result is achieved in one-time-one space dimension (cf. Harrington [1975]). In quantum field theories, the Lagrangian plays the dual role of providing operator equations of motion and of specifying the nature of the ground or vacuum state. An external magnetic field can be readily included in the Lagrangian by the minimal coupling prescription, and the consequent changes in the operator equations of motion are well known. What is not as well understood are the implications of the external magnetic field for the vacuum state of the theory. Since physical quantities are expectation values of operators between states constructed from the vacuum, an alteration of the vacuum can have critical physical consequences.

Sufficiently strong, external magnetic fields can restore a spontaneously broken symmetry by enlarging the symmetry group of the vacuum or create a qualitatively different ground state by energetically favoring a different sector of the theory.

To date, there have been two published contributions that have addressed themselves to symmetry restoration in intense external magnetic fields. We showed (Harrington, Park and Yildiz [1975])that the dynamically broken symmetry in the N-component Thirring model could be restored, and the critical external field was determined. The formalism employed was a combination of the effective potential approach in field theory and the proper time technique (cf. Harrington, Park and Yildiz [1975]). The former was used to search for the dynamically determined ground state while the latter enabled us to include constant external fields of any magnitude. The results obtained in this model are severely limited in their physical applicability due to the peculiarities of the chosen model. The restriction to one space-one time dimension implies no transverse directions so that the external field is electric, not magnetic. Furthermore, the symmetry breaking in the N-component Thirring model appears a $1/N$ expansion and is non-analytic in the coupling. Thus it differs considerably in the symmetry breaking mechanism from the four-dimensional models, such as the Weinberg-Salam model of leptons (cf. Weinberg [1972] and Salam [1969]), the Lee model of CP nonconservation , etc.

The second contribution was that of Salam and Strathdee. Here, four-dimensional models with the symmetry breaking by tachyonic scalar fields were considered. The formalism used to determine the possible existence of symmetry restoration is identical to the two-dimensional approach. However, their work is of a rather of heuristic nature. The most that could be said was that if a critical field exists, then it should be of a specified magnitude. The possible critical fields appear to be very model dependent, ranging from $3x10^6$ gauss for some superweak CP violating theories to 10^{16} gauss for reduction of the Cabibbo angle to zero. Since magnitude fields of the order 10^{12}-10^{14} Gauss may exist in the vicinity of pulsars, the question of symmetry restoration in intense magnetic fields is clearly of physical interest. The main weakness of the Salam-Strathdee approach is their use of the non-renormalizable unitary gauge. Serious doubts must be cast on the calculation of an effective boson mass in the unitary gauge, since this symmetry-determined mass depends critically on the use of renormalization. Salam and Strathdee did not show that higher loop corrections are suppressed for large magnetic fields. In the finite temperature investigations, it was shown that the rise of unitary gauge does not isolate all the leading contributions to the effective boson mass and consequently yields a deceptively incorrect result. Thus it must be concluded that the Salam-Strathdee calculations are to be taken as only indicative of the possible order of magnitude of the critical fields. Studies in four-dimensional models are in consideration by Harrington and Shepard of the University of New Hampshire.

I am grateful to Profsssors Richard Davis and John Lockwood of UNH for financial support. Barry Harrington, Soo Yomg Park and Harvey Shepard deserve my deepest appreciation for many discussions. I am also thankful to Roman Jackiw for many clarifications on the effective potential method.

Fig. 1

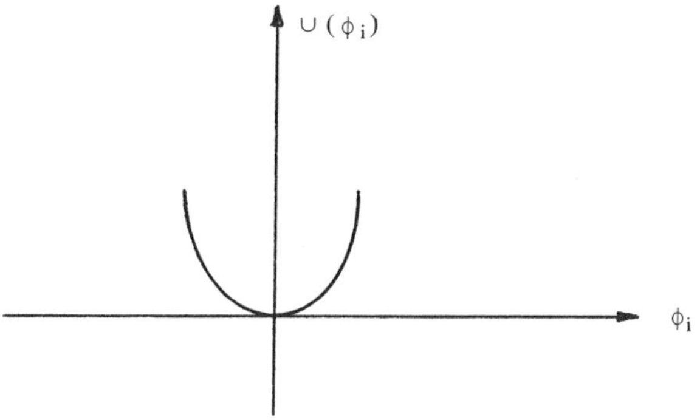

Fig. 2

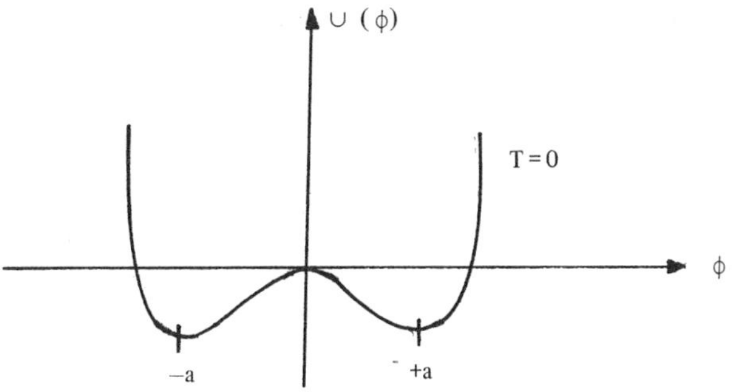

Fig. 3

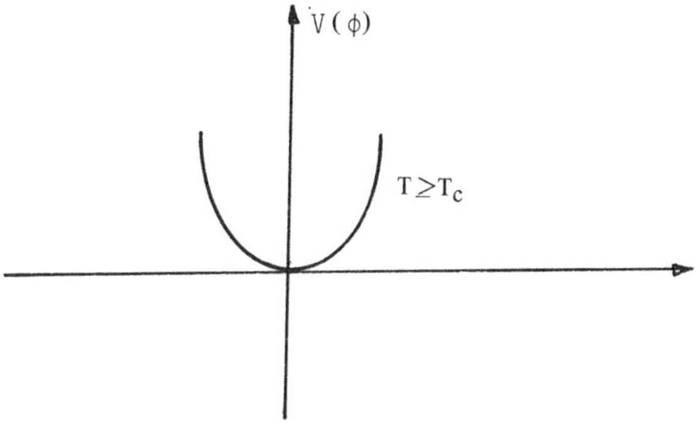

REFERENCES

Bernard, C.W. [1974] *Phys. Rev.* D9, 3312.

Coleman, S. [1973] "*Secret Symmetry*", lectures given at the 1973 International Summer School of Physics "Ettora Majorana."

Dolan, L. and Jackiw, R. [1974] *Phys. Rev.* D9, 3320.

Ginzburg, V.L. and Landau, L.D. [1950] *Zh. Eksp. i Teor. Fiz.* 20, 1064.

Harrington, B.J., Park, S.Y. and Yildiz, A. [1975] *Phys. Rev.* D11, 1472.

Harrington, B.J. and Yildiz, A. [1974] *Phys. Rev. Lett.*33, 324.

Kirzhnits, D.A. and Linde, A.D. [1972] *Phys. Lett.* 42B, 471.

Salam, A. [1969] *Elementary Particle Theory: Relativistic Groups and Analyticity* (Nobel Symposium No.8) edited by N. Svartholm New York:Wiley.

Salam, A. and Strathdee, J. [1974] *Int. Cen. for The.Phys. preprint* IC/74/140.

Weinberg, S. [1972] *Phys. Rev. Lett.* 29, 1698.

Weinberg, S. [1974] *Phys. Rev.* D9, 3357.

ON HIGHER SPIN SUPERFIELDS

D.H. Tchrakian

St. Patrick's College, Maynooth, Ireland
and
Dublin Institute for Advanced Studies

This talk will consist of the conclusions of an investigation by Nilsson and the speaker (Nilsson and Tchrakian [1974]) primarily concerning the construction of superfields, which can be viewed as supermultiplets of Bosons and Fermions or of particles of arbitrary spins. We shall start with a short introduction of the kinematic aspects of supersymmetry covariant fields. Some discussion of the application of this symmetry to the formulation of Lagrangian field theories will indicate the possible applicability of the higher spin superfields.

Problems arising from the non-conservation of Fermion number are not within the scope of this talk, and neither are non-Abelian gauge theoretic supersymmetric Lagrangian theories. These topics are treated by Salam and Strathdee [1974, 1975] as well as the earlier authors Wess and Zumino [1974]. In addition to these references there are two excellent expositions of this subject in the lecture notes of O'Raifeartaigh [1975] and the review article of Corwin, Ne'eman and Sternberg [1975]. Thus all omissions in this lecture are more than adequately dealt with in the above-mentioned references and this lecture could hopefully serve as an introduction to them.

As we have seen very clearly from the preceeding lectures of Professors Barut and Dürr, the role of groups in physics is twofold, the first being that of an invariance or a symmetry, and the second that of a dynamical group which goes beyond the former in that it provides dynamical information on the system. In particular the requirement that the wave function of a system satisfy a conformal-invariant equation gave analytic information on the three-point structure functions or form factors of that system (Barut [1975]) while the same group acting in the role of a covariance group for the non-canonical spinor-fields gave rise to a vector potential which itself is a bilinear construction of the spinor-fields themselves (Dürr [1975]). Naturally all such higher groups should include the Lorentz group as a subgroup.

The supersymmetry group is such a group which contains the Poincaré group, and the dynamical information arising from the invariance of the Lagrangian under its action is that of fixing the coupling constants and interactions of that field theory.

Our presentation here follows that of Salam and Strathdee [1974] using the language of superfields to introduce supersymmetry. The algebra of the infinitesimal generators is

$$[P_\mu, P_\nu] = 0 \tag{1a}$$

$$[S_\alpha, P_\mu] = 0 \tag{1b}$$

$$i[J_{\mu\nu}, P_\lambda] = g_{\mu\lambda}P_\nu - g_{\nu\lambda}P_\mu \tag{1c}$$

$$[S_\alpha, J_{\mu\nu}] = \frac{1}{2}(\sigma_{\mu\nu}S)_\alpha \tag{1d}$$

$$i[J_{\mu\nu},J_{\rho\sigma}] = g_{\mu\rho}J_{\nu\sigma}-g_{\mu\sigma}J_{\nu\rho}-g_{\nu\rho}J_{\mu\sigma}+g_{\nu\sigma}J_{\mu\rho} \tag{1e}$$

$$\{S_\alpha,S_\beta\} = -(\gamma_{\mu\nu}C)_{\alpha\beta}P_\mu \tag{1f}$$

where P_μ and $J_{\mu\nu}$ are the generators of the Poincaré group and $S = (S_a \ \tilde{S}^b)$ are the generators of the supersymmetry transformations labelled by the index $\alpha = a,\dot{b}$; $a,\dot{b} = \pm\frac{1}{2}$, and which further satisfy the Majorana condition

$$S = \varepsilon \ \tilde{S}$$

with ε the two-component antisymmetric symbol (metric spinor).

The commutation relations (C.R) (1b) and (1d) imply respectively that the super-rotations commute with space-time translations and that under Lorentz transformations the supersymmetry (super-rotation) generator behaves like a Dirac spinor, just as P_μ behaves like a four-vector, cf. C.R. (1c).

As for the anticommutation relation (A.C.R.) (1e), it can be thought of as defining the operator S_α as a "square root" of $P.\gamma$, just like $P.\gamma$ itself is the "Square root" of P^2. Also, that (1e) is an A.C.R. means the relations (1) do not form a Lie Algebra (Corwin, Ne'eman and Sternberg [1975]) and hence are not as they stand useful to define supersymmetry-covariant fields (superfields), in the manner that using (1a), (1c) and (1e) one could define Lorentz covariant fields. There however is no obstacle provided we take into account that the group parameter θ corresponding to the generator S_α is itself an anticommuting classical variable. We also require that θ_α be a Majorana spinor

$$\theta_\alpha = (\zeta_a \oplus \eta^{\dot{b}})$$

$$\{\theta_\alpha, \theta_\beta\} = 0 \tag{2}$$

$$\bar{\eta} = \varepsilon\zeta \quad , \quad \bar{\zeta} = \varepsilon\eta. \tag{3}$$

The superfield is taken to be local in x_μ corresponding to P_μ and $J_{\mu\nu}$, and in this classical anticommuting variable θ_α corresponding to S_α. The statements of covariance are

$$U(\Lambda)\Phi^{(A,B)}(x,\theta)U^{-1}(\Lambda) = D^{(A,0)}(\Lambda) \otimes D^{(0,B)}(\Lambda)\Phi^{(A,B)}(\Lambda x, D^{\frac{1}{2}}(\Lambda)\theta) \tag{4}$$

$$U(\phi)\Phi^{(A,B)}(x,\theta)U^{-1}(\phi) = \Phi^{(A,B)}(x+\tfrac{i}{2}\bar{\phi}\gamma_\mu\theta, \quad \theta+\phi). \tag{5}$$

The first of these, (4), follows from the C.R. (1a,c,d,e), and $\Lambda_{\mu\nu}$ is the real Lorentz transformation, $D^{\frac{1}{2}}(\Lambda)$ is the $((\frac{1}{2},0)\oplus(0,\frac{1}{2}))$ representation of the Lorentz group while $D^{(A,0)} \otimes D^{(0,B)}$ is the (A,B) representation. The second one, (5), can be arrived at by considering the group element

$$\exp iP \times \exp i\bar{\theta}S = \Phi(x,\theta)$$

to be a classical scalar superfield, and then operating on it by the super-rotation $\exp i\bar{\phi}S$. Now since θ and ϕ are anticommuting variables, the A.C.R. (1f) becomes the C.R.

$$[\ \bar{\phi}S,\ \bar{\theta}S\ ] = \bar{\phi}\gamma_\mu\theta\ P_\mu, \tag{6}$$

whence by manipulating the matrix exponentials we get

$$(U(\phi)\Phi)(x_\mu,\theta) = \Phi(x_\mu+\tfrac{i}{2}\bar{\phi}\gamma_\mu\theta, \quad \theta+\phi) \tag{5'}$$

which is just (5). In (5) and (6) the meaning of $\bar{\theta}$ is the Dirac conjugate $\bar{\theta} = \theta^*\beta$. Calculating $\delta\Phi$, due to an infinitesimal super-rotation $\delta\theta$ in terms of partial derivatives, we have

$$\delta\Phi = (\overline{\delta\theta}^{\alpha} D_\alpha)\Phi , \tag{7}$$

where D is the super-covariant derivative, and can be read off

$$D_\alpha = \begin{bmatrix} \Delta_a \\ \tilde{\Delta}^{\dot{b}} \end{bmatrix} = \begin{bmatrix} \frac{\partial}{\partial\bar{\eta}} + \frac{i}{2}(\sigma_\mu\eta)\partial_\mu \\ \frac{\partial}{\partial\bar{\zeta}} + \frac{i}{2}(\tilde{\sigma}_\mu\zeta)\partial_\mu \end{bmatrix} , \quad \sigma_\mu = \sigma_o,\sigma_i \tag{8}$$

$$D^*_\alpha = \begin{bmatrix} \overset{*}{\Delta}_{\dot{a}} \\ \overset{*}{\tilde{\Delta}}{}^{b} \end{bmatrix} = \begin{bmatrix} \frac{\partial}{\partial\eta} - \frac{i}{2}(\bar{\eta}\sigma_\mu)\partial_\mu \\ \frac{\partial}{\partial\zeta} - \frac{i}{2}(\bar{\zeta}\tilde{\sigma}_\mu)\partial_\mu \end{bmatrix} , \quad \tilde{\sigma}_\mu = \sigma_o,-\sigma_i \tag{8'}$$

given in terms of two-component notation, for the sake of illustration. These operators (8) and (8') actually do satisfy the A.C.R. (1e) and therefore are differential representations of the operators S_α.

Before going on to decompose the superfield $\Phi^{(A,B)}(x,\theta)$ into its constituent x-local fields of different spins, we make a brief comment on the classical spinor variable θ. The Majorana property (3) seems not to be a necessary condition unless one requires that the supersymmetry induced space-time translation in (5) is real. This may be useful property if one investigates the possibility of introducing a geometry on the (x,θ) space. In any case, as will be seen below, the Majorana restriction tends to make the multiplet structure of the superfields much simpler than otherwise, and this in itself is a very useful quality.

The scalar superfield $\Phi(x,\theta)$ can be expanded in powers of θ whose coefficients are the x-local fields. This expansion terminates after the fourth power of θ, because of the anti-commutativity property (2). If one works with the two-component spinors ζ and η, no more than two powers of either may occur in a non-vanishing term in the expansion. Using the Majorana restriction (3) along with the spin-$\frac{1}{2}$ matrix identity

$$-\sigma_\mu^T = \varepsilon\tilde{\sigma}_\mu\varepsilon \tag{9}$$

one can simplify the covariant bases constructed from $\theta = (\zeta,\eta)$, and many of these bases vanish identically, e.g. $\bar{\theta}\gamma_\mu\theta$. Further reductions are obtained by performing Fierz reshuffles by means of the basic recoupling identity

$$\varepsilon_{ab}\varepsilon_{cd} = \varepsilon_{ac}\varepsilon_{bd} + \varepsilon_{ad}\varepsilon_{cb} \ . \tag{10}$$

For example $(\bar{\theta}\gamma_\mu\gamma_5\theta)(\bar{\theta}\gamma_\nu\gamma_5\theta) = (\bar{\theta}\theta)^2 g_{\mu\nu}$. Complete and systematic tables of such reductions are given in our paper (Nilsson and Tchrakian [1974]).

Accordingly the scalar superfield is decomposed

$$\Phi = A + \bar{\theta}\psi^{(1)} + (\bar{\theta}\theta)F + (\bar{\theta}\theta)(\bar{\theta}\psi^{(2)}) + (\bar{\theta}\theta)^2 D \qquad (11)$$
$$+ (\bar{\theta}\gamma_5\theta)G$$
$$+ (\bar{\theta}i\gamma_\mu\gamma_5\theta)A_\mu$$

where A, F, G and D are spin-0, $\psi^{(1)}$ and $\psi^{(2)}$ spin-$\frac{1}{2}$, and A_μ axial vector fields.

Now we turn our attention to the construction of Lagrangians from such fields before we go on to discussing ways of constructing superfields with higher spin x-local constituents. The following consideration of Lagrangians (Salam and Strathdee [1974]) is given here by way of motivation for the latter.

A Lagrangian density constructed as a power of superfields will be (x,θ)-local, and the invariance of the action

$$\delta\int\mathcal{L}(x,\theta)d^4x = 0$$

due to an infinitesimal super-rotation $\delta\theta$ implies

$$\int(\overline{\delta\theta}D)\mathcal{L}(x,\theta)d^4x = \overline{\delta\theta}\int\frac{\partial\mathcal{L}}{\partial\theta}d^4x = 0$$

neglecting the integral of the x-space surface term. It follows that a supersymmetry invariant Lagrangian must be either independent of θ, or if it is θ-dependent in general, its x-dependent part must be of the form of a space-time divergence.

Now the Lagrangian would be a power of superfields and it is obvious it will be constructed to be a scalar superfield like (11), that it is a supermultiplet terminating with the coefficient of $(\bar{\theta}\theta)^2$. Further, it so happens that calculating the variation due to $\delta\theta$ in the fields A, $\psi^{(1)}$,F,G,$\psi^{(2)}$,A_μ

and D of (11) by means of (7), it is found that the variation δD is itself a pure space-time divergence

$$\delta D = -2\partial_\mu \overline{\delta\theta}\gamma_\mu \psi^{(2)}$$

and hence by the above reasoning D is the suitable candidate for a Lagrangian density.

The above construction may be further simplified by subjecting the superfields to further conditions, and for scalar superfields we can require the property of reality

$$\Phi(x,\theta) = \Phi^*(x,\theta)$$

which causes all x-local Bose fields to be real and all Fermi fields to be of Majorana type. Another condition is the chirality condition

$$\Delta_a \Phi_- = 0 \qquad (12a)$$

$$\tilde{\Delta}^{\dot{b}} \Phi_+ = 0. \qquad (12b)$$

The superfields Φ_+ are called chiral superfields and their spin content is less than that of Φ. They are in fact the superfields that yield the Wess-Zumino Lagrangian (Zumino [1972]; O'Raifeartaigh [1975]).

Imposing (12a) on the expansion (11) yields the following conditions, using the notation $\psi = (\Phi \oplus \chi)$,

$$\phi^{(1)} = 0, \qquad \chi^{(2)} = 0, \qquad \phi^{(2)} = -\tfrac{i}{2}\sigma_\mu \partial_\mu \chi^{(1)} \qquad (13)$$

$$F+G=0, \qquad \partial_\mu A_\nu - \partial_\nu A_\mu = 0, \qquad A_\mu = -\tfrac{1}{2}\partial_\mu A, \qquad D = -\tfrac{i}{8}\partial_\mu A_\mu$$

which yields the chiral scalar superfield

$$\Phi_-=A+\bar{\zeta}\chi^{(1)}+(\bar{\zeta}\eta)(F-G)+\frac{i}{4}(\bar{\zeta}\eta)\bar{\eta}\sigma_\mu\partial_\mu\chi^{(1)}-\frac{1}{16}(\bar{\eta}\zeta)(\bar{\zeta}\eta)\partial^2A-\frac{i}{2}(\bar{\eta}\sigma_\mu\eta)\partial_\mu A \quad (14)$$

and a similar expression for Φ_+. The chiral superfield therefore contains only one spin-$\frac{1}{2}$ Majorana field and two scalar fields A and (F-G). Moreover, we have the further facility that the superfield itself incorporates the kinetic terms of the x-local fields, and therefore products of chiral superfields give rise to Lagrangian densities without further differentiation (O'Raifeartaigh [1975]).

We see therefore that if we want to construct, following the above prescription, a supersymmetric theory of particles with spin higher than one half, we will need higher spin superfields.

The highest spin occurring in the unrestricted scalar superfield is spin-1, and it corresponds to the basis $\bar{\theta}i\gamma_\mu\gamma_5\theta$. Now to obtain higher spins there appear to be three obvious constructions for the superfield:

(i) To define a scalar superfield local in x and a 2(2j+1) component classical anticommuting variable $\theta^{(j)}$. In this case one would have non-vanishing θ-basis that include up to 2(2j+1) factors of θ, and hence the highest spin field contained has spin-(2j+1). Of course if this spinor is subjected to the Majorana condition, the highest spin will be lower.

The defect of this method is that generalizing the commutation relations (1d) and (1f) we would naturally use, respectively as structure constants

$$J_{\mu\nu}((j,0)\oplus(0,j) \;\&\; \gamma_{\mu_1 \cdot\cdot \mu_{2j}} = (-i)^{2j} \begin{bmatrix} 0 & \sigma_{(\mu)}(j,j) \\ \tilde{\sigma}_{(\mu)}(j,j) & 0 \end{bmatrix}$$

The latter would need 2j powers of the translation generator P_μ to contract all the space-time indices on $\gamma_{(\mu)}$, and this would in turn not result into the simple (required) covariance condition (5) and (5').

(ii) If it is desired to have a Lorentz scalar superfield, one could incorporate higher spins by simply defining a superfield that depends on x and n classical anticommuting four component spinors $\theta^{(1)}, \theta^{(2)}, \ldots, \theta^{(n)}$. In this case the definition of supersymmetry itself must be modified such that the infinitesimal generators $S^{(i)}$ (i=1,...,n) corresponding to each group parameter $^{(i)}$ satisfying the following A.C.R.s

$$\{S_\alpha^{(i)}, S_\beta^{(j)}\} = 0 \qquad , \; i\neq j \tag{15}$$

$$\{S_\alpha^{(i)}, S_\beta^{(j)}\} = -(\gamma_\mu C)_{\alpha\beta} P_\mu \qquad , \; i=j. \tag{16}$$

and $S^{(i)}$ satisfy (1b) and (1d) also- for all i. In this case the space-time translation induced by a super-rotation $(\phi^{(1)},\ldots,\phi^{(n)})$ is

$$x_\mu \rightarrow x_\mu + \frac{i}{2} \sum_i^n \bar{\phi}^{(i)} \gamma_\mu \theta^{(i)} \tag{17}$$

which generalizes (5) and (5').

This is a consistent procedure, and for a superfield depending on n spinors, the maximum power of the spinor basis is 4n which could describe a particle of spin-2n. If in addition these spinors satisfy the Majorana condition, and (17) becomes real, the highest spin that can be described is spin-n.

The weakness of this procedure however is that the simplification ensuing from the identity of the spinors in the one-spinor case is no more forthcoming because of the use of distinct $\theta^{(i)}$. This tends to make the supermultiplet structure unwieldy.

(iii) This is the simplest and most satisfactory method, and consists of the use of Lorentz covariant superfields $\Phi(A,B)$, which was anticipated in the statements of covariance in (4) and (5), and are implicitly present in the literature (Salam and Strathdee [1974]; O'Raifeartaigh [1975]).

The covariant superfields have decompositions of the type (11), with the fourth power of θ terminating the supermultiplet series. The θ-basis to be used, and the various identities for reducing them with respect to the representations of the Lorentz group are given in detail in our paper (Nilsson and Tchrakian [1974]), and we shall omit those technicalities here. We only note that the highest spin field contained in the (A,B) Lorentz covariant superfield is $(|A+B|+1)$, while the lowest is $(|A-B|=1)$. To illustrate the actual decomposition of such superfields we list below two examples, the Vector $\Phi_\mu(x,\theta)$ and the Dirac $\psi_\alpha(x,\theta)$ superfields, each transforming respectively according to the representations $(\frac{1}{2},\frac{1}{2})$ and $((\frac{1}{2},0)\oplus(0,\frac{1}{2}))$ of the Lorentz group

$$\begin{aligned}\Phi_\mu = V_\mu^{(1)} + \bar\theta\psi_\mu^{(1)} &+ (\bar\theta\theta)V_\mu^{(2)} + (\bar\theta\theta)\bar\theta\psi_\mu^{(2)} + (\bar\theta\theta)^2 V_\mu^{(3)} \\ + \bar\theta i\gamma_\mu\psi^{(1)} &+ (\bar\theta\gamma_5\theta)A_\mu + (\bar\theta\theta)\bar\theta i\gamma_\mu\psi^{(2)} \\ &+ (\bar\theta i\gamma_\mu\gamma_5\theta)P \\ &+ (\bar\theta i\gamma_\nu\gamma_5\theta)H_{\mu\nu}\end{aligned} \tag{18}$$

A "Lorentz" condition may be used to reduce the multiplicity of (18). Thus the condition $\partial_\mu\Phi_\mu = 0$ gives rise to the following restrictions on the x-local component fields

$$\begin{aligned}&\partial_\mu\psi_\mu^{(i)} + \psi^{(i)} = 0, \quad i=1,2 \\ &\partial_\mu A_\mu = 0, \qquad \partial_\mu V_\mu^{(i)} = 0, \quad i=1,2,3 \\ &\partial_\mu P + \partial_\nu H_{\mu\nu} = 0.\end{aligned} \tag{19}$$

It can be seen from (19) that the condition $\partial_\mu\Phi_\mu = 0$ eliminates two spin-$\frac{1}{2}$ and three spin-0 fields from the supermultiplet (18), but it is to be noted that it does not eliminate the highest spin field, in contrast to the chirality condition (12) on the scalar superfield.

Next we write down the decomposition of the Dirac superfield

$$\begin{aligned}\Psi = \psi^{(1)} + \theta\, S^{(1)} &+ (\bar\theta\theta)\psi^{(2)} + (\bar\theta\theta)\theta\, S^{(2)} + (\bar\theta\theta)^2\psi^{(4)} \\ +\gamma_5\theta\, P^{(1)} &+ (\bar\theta\gamma_5\theta)\psi^{(3)} + (\bar\theta\gamma_5\theta)\theta\, P^{(2)} \\ +i\gamma_\mu\theta\, V^{(1)} &+ (\bar\theta i\gamma_\mu\gamma_5\theta)\psi_\mu + (\bar\theta\theta)i\gamma_\mu\theta\, V^{(2)} \\ +i\gamma_\mu\gamma_5\theta\, A^{(1)} &\qquad + (\bar\theta\gamma_5\theta)i\gamma_\mu\theta\, A^{(2)} \\ +\gamma_{[\mu}\gamma_{\nu]}\theta\, F_{\mu\nu}^{(1)} &\qquad + (\bar\theta\theta)\gamma_{[\mu}\gamma_{\nu]}\theta\, F_{\mu\nu}^{(2)} \\ +\gamma_{[\mu}\gamma_{\nu]}\gamma_5\theta\, G_{\mu\nu}^{(1)} &\qquad + (\bar\theta\gamma_5\theta)\gamma_{[\mu}\gamma_{\nu]}\theta\, G_{\mu\nu}^{(2)}\end{aligned} \tag{20}$$

Just as the "Lorentz" condition eliminated some of the fields in the supermultiplet (18), the free Dirac equation enforced on (20) results in some eliminations arising from free equations on the component fields of (20). Thus, as a result of applying $(\gamma.\partial - m)\psi = 0$, we end up with the following relations, for i=1,2.

$$mS^{(i)} = i\partial_\mu V_\mu^{(i)} \qquad mP^{(i)} = i\partial_\mu A_\mu^{(i)} \tag{21}$$
$$imV_\mu^{(i)} = \partial_\mu S^{(i)} + 4\partial_\nu F_{\mu\nu}^{(i)} \qquad imA_\mu^{(i)} = \partial_\mu P^{(i)} + 4\partial_\nu G_{\mu\nu}^{(i)}$$
$$mF_{\mu\nu}^{(i)} = \frac{i}{4}(\partial_\mu V_\nu^{(i)} - \partial_\nu V_\mu^{(i)}) \qquad mG_{\mu\nu}^{(i)} = \frac{i}{4}(\partial_\mu A_\nu^{(i)} - \partial_\nu A_\mu^{(i)})$$

in addition to free Dirac equations for all the Fermion fields in (20). The nomenclature in (20) and (21) is that, S is scalar, P Pseudoscalar, V_μ vector, A_μ axial vector, $F_{\mu\nu}$ is 1^- and $G_{\mu\nu}$ is 1^+. As in the case of conditions (19), the component field of highest spin is not eliminated.

More useful in the construction of Lagrangians are chiral conditions of the type (12), so we shall conclude by an example which illustrates the procedure for constructing covariant chirality superfields. The example is the Dirac superfield (not subject to the "Dirac equation" or (21)). Two sets of conditions are enforced on (20). They are respectively

$$D^\alpha \psi_\alpha = 0 \tag{22}$$

$$\overset{\star}{\Delta}{}^a \Phi_a = 0$$
$$\overset{\star}{\Delta}_{\dot{b}} X^{\dot{b}} = 0, \tag{23}$$

where we have denoted $\Psi = (\Phi_a \oplus X^b)$. The condition (22) leads to

$$S^{(1)} = 0, \qquad \partial_\mu V_\mu^{(2)} = 0$$
$$S^{(2)} = \frac{i}{4}\partial_\mu V_\mu^{(1)}, \qquad P^{(2)} = -\frac{i}{4}\partial_\mu A_\mu^{(1)} \qquad (24)$$
$$A_\mu^{(2)} = -\frac{i}{4}(\partial_\mu P^{(1)} + 4\partial_\nu(\tilde{F}_{\mu\nu} - G_{\mu\nu}))$$

$$\psi^{(3)} = -\gamma_5\psi^{(2)} - \frac{i}{2}\gamma_\mu\psi_\mu - \frac{1}{4}\gamma_\mu\partial_\mu\psi^{(1)} \qquad (24')$$
$$\psi^{(4)} = -\frac{1}{16}\partial^2\psi^{(1)} + \frac{i}{8}\gamma_{[\mu}\gamma_{\nu]}\gamma_5\partial_\mu\psi_\nu$$

which eliminate four spin-0, one axial vector and two Dirac fields, but not the highest spin field, namely the Rarita-Schwinger.

More restrictive conditions arise from (23). They are

$$S^{(1)} = P^{(1)} = 0, \qquad S^{(2)} = \frac{i}{4}\partial_\mu V_\mu^{(1)}, \qquad P^{(2)} = -\frac{i}{4}\partial_\mu A_\mu^{(1)}$$
$$V_\mu^{(2)} = i\partial_\nu(F_{\mu\nu}^{(1)} - \tilde{G}_{\mu\nu}^{(1)}) \qquad (25)$$
$$A_\mu^{(2)} = -i\partial_\nu(\tilde{F}_{\mu\nu}^{(1)} - G_{\mu\nu}^{(1)})$$

$$\psi^{(3)} = -\gamma_5\psi^{(2)}$$
$$\psi_\mu = -\frac{i}{2}\partial_\mu\gamma_5\psi^{(1)} \qquad (25')$$
$$\psi^{(4)} = -\frac{1}{16}\partial^2\psi^{(1)}$$

which eliminates the Rarita-Schwinger field, and is similar to (13), the chirality condition for the scalar superfield, in the sense that it eliminates the highest spin field of the supermultiplet.

As can be seen from (25), a Lorentz invariant bilinear product of Dirac superfields subject to (23) will contain terms that would be candidates for the kinetic part of the Lagrangian density. Denoting $\bar{\Psi} = \Psi^* \beta$, the bilinear product is $\bar{\Psi}\Psi$ and a possible interaction term is $g(\bar{\Psi}\Psi)\Phi$, and of course their sum is a scalar superfield. Hence picking out of it the coefficient of $(\bar{\theta}\theta)^2$, one ends up with a suitable supersymmetric Lagrangian density.

Finally we comment that in both (24) and (25) one has the unusual type of term

$$\psi^{(4)} \sim \partial^2 \psi^{(1)} + \ldots$$

which would result into Boson type kinetic terms for a Fermion field. This is a common occurrence for covariant chiral superfields that have Fermionic Lorentz group index, that is, A-B is half integral. It is not a surprising defect in a supersymmetric theory (Salam and Strathdee [1975]), and can be avoided if only superfields $\Phi(A,B)$ with integral values for A-B are used.

REFERENCES

Barut, A.O. [1975] in the Proceedings of the International Symposium in *Mathematical Physics*, Boğaziçi University, Istanbul.

Corwin, L., Ne'eman, Y. and Sternberg, S. [1975], *Rev. Mod. Phys.* 47, 573.

Dürr, H.P., in the Proceedings of the International Symposium in *Mathematical Physics*, Boğaziçi University, Istanbul [1975].

Nilsson, J.S. and Tchrakian, D.H. [1974], Goteborg preprint 74-44, To be published in *Physica Scripta* (Sweden).

O'Raifeartaigh, L. [1975] *Lecture Notes on Supersymmetry*, Communications of the Dublin Institute for Advanced Studies.

Salam, A. and Strathdee, J. [1974] *Trieste Report* IC/74/42.

Salam, A. and Strathdee, J. [1975] *Trieste Report* IC/75/49.

Zumino, B. [1974] in the Proceedings of the XVII International Conference on *High Energy Physics*, London (CERN rep. TH-1901).

CP NONCONSERVATION IN TWO QUARK DOUBLET MODELS

Burhan Cahit Ünal*

Institut für Theoretische Physik
der Universität Heidelberg

ABSTRACT

T.D. Lee mechanism of simultaneous CP and gauge symmetry breaking is applied to GIM and Achiman models.

1. PHASE CONVENTION FOR K^o, $\bar{K}^o$ SYSTEM

a) Under charge conjugation

$$\begin{aligned} CK^o C^{-1} &= -\bar{K}^o \\ C\bar{K}^o C^{-1} &= -K^o \end{aligned} \qquad (1.1)$$

as K^o and $\bar{K}^o$ are pseudoscalar, we have under CP

$$\begin{aligned} CPK^o (CP)^{-1} &= \bar{K}^o \\ CP\bar{K}^o (CP)^{-1} &= K^o \end{aligned} \qquad (1.2)$$

b) Transformation under time reversal is obtained from pseudoscalar coupling

$$\mathcal{L} = ig\tilde{n}\gamma_5 \lambda K^o + ig\tilde{\lambda}\gamma_5 n\bar{K}^o$$

as follows

$$\begin{aligned} \mathcal{T}\mathcal{L}(\vec{x},t)\mathcal{T}^{-1} &= -ign^\dagger(\vec{x},-t)T^{-1}(\gamma^o\gamma_5)^* T\lambda(\vec{x},-t)\mathcal{T}K^o\mathcal{T}^{-1} \\ &= -ig\tilde{n}(\vec{x},-t)\gamma_5\lambda(\vec{x},-t)\mathcal{T}K^o\mathcal{T}^{-1} + \text{h.c.} \\ &= \mathcal{L}(\vec{x},-t) \end{aligned}$$

* Alexander von Humboldt Stiftiung's fellow. On leave of absence from the University of Ankara.

if

$$\mathcal{T} K^{o}(\vec{x},t)\mathcal{T}^{-1} = -K^{o}(\vec{x},-t). \tag{1.3}$$

Thus under CPT we shall have

$$\begin{aligned} \text{CPT } K^{o}(\text{CPT})^{-1} &= -\bar{K}^{o} \\ \text{CPT } \bar{K}^{o}(\text{CPT})^{-1} &= -K^{o} \end{aligned} \tag{1.4}$$

2. PROPAGATOR APPROACH

In the propagator approach (cf. Sachs [1965]) the masses and lifetimes of the particles are expressed in terms of properties of the propagator. The boson propagator is

$$\Delta'(k^2) = \lim_{Z\to k^2+i\varepsilon} \frac{1}{M(Z) - Z} \tag{2.1}$$

where $M(Z)$ is the renormalized mass term to which all the proper self-energy graphs are added. For an unstable particle there will be a pole in the propagator in the second sheet corresponding to the solution of the equation

$$M(Z_o) = Z_o \tag{2.2}$$

The mass and the lifetime are related to Z_o through the expression

$$Z_o = (m - i\,\frac{\gamma}{2})^2 \tag{2.3}$$

K^o system is a twofold degenerate system under strong and electromagnetic interactions. The weak interactions introduce the self-energy corrections and mixes the two particle states. The propagator and therefore $M(z)$ becomes now a 2×2 matrix. m_o being the "bare" mass, i.e. the mass with all weak interactions turned off, the general form of M will be

$$M = \begin{bmatrix} m_o^2 + w & p^2 \\ q^2 & m_o^2 + w' \end{bmatrix} \begin{matrix} K^o \\ \bar{K}^o \end{matrix} \qquad \text{columns: } K^o,\ \bar{K}^o \tag{2.4}$$

under CPT invariance $w = w'$. Introducing $r = q/p$ the poles close to the physical sheet can be expressed as

$$Z_S = m_o^2 + w + pq \quad \text{(short lifetime)}$$
$$Z_L = m_o^2 + w - pq \quad \text{(long lifetime)} \qquad (2.5)$$

and the eigenstates are

$$|K_S\rangle = |K^o\rangle + r|\bar{K}^o\rangle = (1+r)|K_1^o\rangle + (1-r)|K_2^o\rangle$$
$$|K_L\rangle = |K^o\rangle - r|\bar{K}^o\rangle = (1+r)|K_2^o\rangle + (1-r)|K_1^o\rangle \qquad (2.6)$$

The mass and lifetime difference is

$$Z_S - Z_L = 2pq = 2(M_{K\bar{K}} M_{\bar{K}K})^{1/2} \qquad (2.7)$$

and the CP violating coefficient is

$$\varepsilon = \frac{1-r}{1+r} = \frac{p^2-q^2}{(p+q)^2} = \frac{M_{K\bar{K}} - M_{\bar{K}K}}{M_{K\bar{K}} + M_{\bar{K}K} + 2(M_{K\bar{K}} M_{\bar{K}K})^{1/2}}$$
$$= \frac{M_{K\bar{K}} - M_{\bar{K}K}}{Z_S - Z_L + M_{K\bar{K}} + M_{\bar{K}K}} \qquad (2.8)$$

These expressions are the most general and M matrix is not hermitian in the propagator approach (cf. Jacob and Sachs [1961]).

The M matrix will have CP invariant and CP non-invariant parts:

$$M = M^{(+)} + M^{(-)}$$

such that

$$CP\, M^{(+)}_{K\bar{K}}\, (CP)^{-1} = M^{(+)}_{\bar{K}K} \neq (M^{(+)}_{K\bar{K}})^*$$

$$CP\, M^{(-)}_{K\bar{K}}\, (CP)^{-1} = -M^{(-)}_{\bar{K}K} \neq -(M^{(-)}_{K\bar{K}})^* \qquad (2.9)$$

ε becomes, in terms of these two parts

$$\varepsilon = \frac{2\, M^{(-)}_{K\bar{K}}}{Z_S - Z_L + 2\, M^{(+)}_{K\bar{K}}}$$

In the miliweak and superweak theories $Z_S - Z_L$ comes mainly from CP conserving part, then from eq.(2.7)

$$Z_S - Z_L = 2\, M^{(+)}_{K\bar{K}}$$

and

$$\varepsilon = \frac{M^{(-)}_{K\bar{K}}}{2\, M^{(+)}_{K\bar{K}}} \qquad (2.10)$$

3. EXTENSION OF WEINBERG-SALAM MODEL TO HADRONS

In the extensions of Weinberg-Salam model to hadrons one can take two SU(2) doublets of quarks of Y = 1

$$N_L^1 = \begin{pmatrix} p_1 \\ n^c \end{pmatrix}_L , \quad N_L^2 = \begin{pmatrix} p_2 \\ \lambda^c \end{pmatrix}_L \tag{3.1}$$

where p_2 is the charmed quark of GIM (cf. Glashow, Illiopoulos, Maiani [1970]) or colored quark of Achiman (cf. Achiman [1973]) and n^c and λ^c are Cabibbo rotated fields

$$\begin{aligned} n^c &= n_1 \cos\theta + \lambda_1 \sin\theta \\ \lambda^c &= -n_1 \sin\theta + \lambda_1 \cos\theta \end{aligned} \tag{3.2}$$

SU(2) singlets are p_{1R}, p_{2R} of Y = 2 and n_R, λ_R of Y = 0.

As gauge fields, we have corresponding to SU(2) gauge group the usual three A_μ^j vector bosons and corresponding to U(1) the B_μ.

The Higgs scalars are also introduced (similar to fermions) as two doublets of Y = 1

$$\phi_1 = \begin{pmatrix} \phi_1^+ \\ \phi_1^o \end{pmatrix}, \quad \phi_2 = \begin{pmatrix} \phi_2^+ \\ \phi_2^o \end{pmatrix} \tag{3.3}$$

their adjoints of Y = -1 are

$$\tilde{\phi}_1 = i\sigma_2 \phi_1^* = \begin{pmatrix} \phi_1^{o*} \\ -\phi_1^- \end{pmatrix}, \quad \tilde{\phi}_2 = i\sigma_2 \phi_2^* = \begin{pmatrix} \phi_2^{o*} \\ -\phi_2^- \end{pmatrix} \tag{3.4}$$

Their coupling to the gauge fields is completely determined by the requirements of gauge invariance. Since T is an antiunitary operator, we can always choose the phase of ϕ_i such that

$$T\phi_i T^{-1} = \phi_i \tag{3.5}$$

The T and gauge invariant Lagrangian density (cf. Abers and Lee [1973]) is

$$L = L(W) + L(W,\phi) + L(h,\phi) + L(h,W) \tag{3.6}$$

where

$$L(W) = -\frac{1}{4} F^i_{\mu\nu} F^{i\mu\nu} - \frac{1}{4} B_{\mu\nu} B^{\mu\nu} \tag{3.7}$$

with

$$F^i_{\mu\nu} = \partial_\mu A^i_\nu - \partial_\nu A^i_\mu + g\varepsilon^{ijk} A^j_\mu A^k_\nu \tag{3.8}$$

$$B_{\mu\nu} = \partial_\mu B_\nu - \partial_\nu B_\mu$$

$$L(W,\phi) = (D_\mu \phi_i)^+ (D^\mu \phi_i) + V(\phi_i) \tag{3.9}$$

with

$$D_\mu = \partial_\mu - i\frac{1}{2} g' B_\mu - i\frac{1}{2} g\tau^j A^j_\mu \tag{3.10}$$

$$\begin{aligned} V(\phi_i) = {} & -\lambda_1 \phi_1^\dagger \phi_1 - \lambda_2 \phi_2^\dagger \phi_2 + A(\phi_1^\dagger \phi_1)^2 + B(\phi_2^\dagger \phi_2)^2 \\ & + C(\phi_1^+ \phi_1)(\phi_2^+ \phi_2) + \bar{C}(\phi_1^+ \phi_2)(\phi_2^+ \phi_1) \\ & + \frac{1}{2}[(\phi_1^\dagger \phi_2)(D\phi_1^\dagger \phi_2 + E\phi_1^\dagger \phi_1 + F\phi_2^\dagger \phi_2 + \text{h.c.}] \end{aligned} \tag{3.11}$$

the nine real constants in V assure the T invariance.

$$\begin{aligned} L(h,\phi) = {} & E^i_j \tilde{N}^i_L \tilde{\phi}_j p_{1R} + F^i_j \tilde{N}^i_L \tilde{\phi}_j p_{2R} \\ & + G^i_j \tilde{N}^i_L \phi_j n_R + H^i_j \tilde{N}^i_L \phi_j \lambda_R + \text{h.c.} \end{aligned} \tag{3.12}$$

with $(E,F,G,H)^{i}_{j}$ all real, $i,j = 1,2$.

$$L(h,\ W) = \tilde{N}^{i}_{L}\ i\gamma^{\mu} D_{\mu} N^{i}_{L} + p_{iR}\, i\gamma^{\mu} D^{s}_{\mu} p_{iR} \tag{3.13}$$

with

$$\begin{aligned} D_{\mu} &= \partial_{\mu} - \frac{i}{2}\, g' B_{\mu} - \frac{i}{2} g \tau^{j} A^{j}_{\mu} \\ D^{s}_{\mu} &= \partial_{\mu} - i g' B_{\mu} \end{aligned} \tag{3.14}$$

4. T.D. LEE MECHANISM OF SPONTANEOUS SYMMETRY BREAKING

The spontaneous T-violation mechanism is introduced by assuming a T-invariant potential energy $V(\phi_i)$, Eq.(3.11) which has a minimum at the c number point

$$\begin{pmatrix} \phi_1^+ \\ \phi_1^o \end{pmatrix} = \frac{1}{\sqrt{2}} \begin{pmatrix} 0 \\ \rho_1 e^{i\varphi} \end{pmatrix}, \begin{pmatrix} \phi_2^+ \\ \phi_2^o \end{pmatrix} = \frac{1}{\sqrt{2}} \begin{pmatrix} 0 \\ \rho_2 \end{pmatrix} \tag{4.1}$$

if λ_1 and/or $\lambda_2 > 0$, $D > 0$ and $D > \bar{c}$. In the tree approximation the numbers in Eq.(4.1) are interpreted as vacuum expectation values of the corresponding fields. Such that the two symmetries are simultaneously broken. These numbers do not satisfy either Eq.(3.5) and the gauge symmetry SU(2) x U(1) of the Lagrangian. Since one of the four generators, namely τ_3+1 is not broken, other three τ_1, τ_2, τ_3-1 are broken. If only T were violated, while the gauge symmetry were not, one could obtain T or CP violation of an arbitrary magnitude. But here, the spontaneous T violation being coupled with the spontaneous gauge symmetry violation, the arbitrariness can be limited.

As three generators of the gauge group are broken, there will be three massless Goldstone bosons; these are

$$G^o = \frac{1}{\sqrt{\rho_1^2 + \rho_2^2}} (\rho_1 I_1 + \rho_2 I_2) \tag{4.2}$$

$$G^{\pm} = \frac{1}{\sqrt{\rho_1^2 + \rho_2^2}} (\rho_1 \phi_1^{\pm} e^{\pm i\varphi} + \rho_2 \phi_2^{\pm}) \tag{4.3}$$

from eight real fields it remains five massive bosons: three vibrational modes corresponding to the quantum fluctuations of the triangle around the minimum point are

$$\begin{bmatrix} t_1 \\ t_2 \\ t_3 \end{bmatrix} = U \begin{bmatrix} R_1 \\ R_2 \\ I \end{bmatrix} \quad \text{where} \quad I = \frac{1}{\sqrt{\rho_1^2 + \rho_2^2}} (\rho_2 I_1 - \rho_1 I_2) \tag{4.4}$$

of masses m_1, m_2, m_3 and two charged bosons

$$H^{\pm} = \frac{1}{\sqrt{\rho_1^2 + \rho_2^2}} (\rho_2 \phi_1^{\pm} e^{\pm i\varphi} - \rho_1 \phi_2^{\pm}) \tag{4.5}$$

of masses

$$m_H^2 = \frac{1}{2} (D - \bar{C})(\rho_1^2 + \rho_2^2) \tag{4.6}$$

We notice that in the Lagrangian only Higgs scalars have mass terms , all other particles gain their masses by the breaking of the gauge symmetry. We are interested mainly in the quark masses, while the following combinations of the gauge fields

$$W_\mu^{\pm} = \frac{1}{\sqrt{2}} (A_\mu^1 \pm iA_\mu^2), \qquad Z_\mu = \frac{g' B_\mu - g A_\mu^3}{\sqrt{g^2 + g'^2}}, \qquad A_\mu = \frac{g B_\mu + g' A_\mu^3}{\sqrt{g^2 + g'^2}} \tag{4.7}$$

have respectively the following masses:

$$m_W = \frac{g}{2}\sqrt{\rho_1^2 + \rho_2^2}\,, \quad m_Z = \frac{1}{2}\sqrt{g^2+g'^2}\,\sqrt{\rho_1^2+\rho_2^2}\,, \quad m_A = 0 \tag{4.8}$$

5. QUARK MASSES

In the Lagrangian $L(h,\phi)$, Eq.(3.12), replacing the coefficients by the following values

$E_1^2 = E_2^2 = F_1^1 = F_2^1 = 0$ (to suppress the cross terms such as $\tilde{p}_{2L}\, p_{1R}$ and $\tilde{p}_{1L}\, p_{2R}$)

$$E_i^1 = k_i^{p_1} \;;\quad F_i^2 = k_i^{p_2}$$

$$G_i^1 = k_i^n \cos\theta;\quad G_i^2 = -k_i^n \sin\theta \tag{5.1}$$

$$H_i^1 = k_i^\lambda \sin\theta;\quad H_i^2 = k_i^\lambda \cos\theta$$

and combining the k_i^h 's as follows

$$k_1^h\, \rho_1 e^{i\varphi} + k_2^h\, \rho_2 = \sqrt{2}\, m_h e^{i\alpha_h} \quad \text{with} \quad h = p_1, p_2, n, \lambda \tag{5.2}$$

We obtain

$$L(h,\phi) = m_1 \tilde{p}_{1L} p_{1R} e^{-i\alpha_{p_1}} + m_2 \tilde{p}_{2L} p_{2R} e^{-i\alpha p_2}$$

$$+ m_n \tilde{n}_L n_R e^{i\alpha_n} + m_\lambda \tilde{\lambda}_L \lambda_R e^{i\alpha_\lambda} + h.c.$$

To have a real diagonal mass we must transform the right-handed quarks as

$$p_{iR} e^{-i\alpha_{p_i}} = (\psi_{p_i})_R \qquad i = 1,2$$

$$n_R e^{i\alpha_n} = (\psi_n)_R \tag{5.3}$$

$$\lambda_R e^{i\alpha_\lambda} = (\psi_\lambda)_R$$

Lagrangian becomes

$$L(h,\phi) = m_1 \tilde{\psi}_{p_1} \psi_{p_1} + m_2 \tilde{\psi}_{p_2} \psi_{p_2} + m_n \tilde{\psi}_n \psi_n + m_\lambda \tilde{\psi}_\lambda \psi_\lambda \quad (5.4)$$

We notice that the coupling between Higgs scalars and quarks and their vacuum expectation values determine hadron masses m^h as in Eq.(5.2).

Using the expressions Eq.(5.1) in the Lagrangian in Eq.(3.12) we shall obtain the interaction terms between different components of Higgs scalars and hadrons. We are mainly interested in the coupling terms $NNW^{\pm}$, $NNG^{\pm}$, $NNH^{\pm}$

$$L_{NNW^+} = \frac{m_w W^+_\mu}{\sqrt{2(\rho_1^2+\rho_2^2)}} [\tilde{p}_1 \gamma^\mu (1-\gamma_5) n \cos\theta + \tilde{p}_1 \gamma^\mu (1-\gamma_5) \lambda \sin\theta$$

$$- \tilde{p}_2 \gamma^\mu (1-\gamma_5) n \sin\theta + \tilde{p}_2 \gamma^\mu (1-\gamma_5) \lambda \cos\theta]$$

$$(5.5)$$

$$L_{NNG^+} = \frac{G^+}{\sqrt{2(\rho_1^2+\rho_2^2)}} \{ -\tilde{p}_1 [m_1 (1-\gamma_5) - m_n (1+\gamma_5)] n \cos\theta -$$

$$- \tilde{p}_1 [m_1 (1-\gamma_5) - m_\lambda (1+\gamma_5)] \lambda \sin\theta$$

$$+ \tilde{p}_2 [m_2 (1-\gamma_5) - m_n (1+\gamma_5)] n \sin\theta -$$

$$- \tilde{p}_2 [m_2 (1-\gamma_5) - m_\lambda (1+\gamma_5)] \lambda \cos\theta \}$$

$$L_{NNH^+} = \frac{(\rho_2/\rho_1)H^+}{\sqrt{2(\rho_1^2+\rho_2^2)}} \{-\tilde{p}_1[\kappa_1(1-\gamma_5)-\kappa_n(1+\gamma_5)]n\cos\theta$$

$$-\tilde{p}_1[\kappa_1(1-\gamma_5)-\kappa_\lambda(1+\gamma_5)]\lambda\sin\theta$$

$$+\tilde{p}_2[\kappa_2(1-\gamma_5)-\kappa_n(1+\gamma_5)]n\sin\theta$$

$$-\tilde{p}_2[\kappa_2(1-\gamma_5)-\kappa_\lambda(1+\gamma_5)]\lambda\cos\theta\}$$

where

$$\kappa_h = m_h - \frac{\rho_1^2+\rho_2^2}{\sqrt{2}\,\rho_2} k_2^h e^{-i\alpha_h} \qquad (5.6)$$

It is easy to verify that L_{NNW} and L_{NNG} are CP conserving, while L_{NNH} is CP non-conserving.

6. ε PARAMETER

We have to compute CP conserving and CP non-conserving scattering matrix elements. The lowest order diagrams for $\Delta Y = \pm 2$ process

$$n\bar{\lambda} \rightleftarrows W^+ W^- \rightleftarrows \lambda \bar{n} \qquad (6.1)$$

are fourth order ones.

For T conserving scattering we have to compute the two diagrams in Fig.1. While for T nonconserving scattering the diagrams in Fig.2 have to be computed. We have used R_ξ gauge , such that the propagators are

$$\text{for W boson} \qquad \Delta^W_{\mu\nu}(q;\xi) = \frac{1}{q^2 - m_W^2}\left[-g_{\mu\nu} + \frac{(1-1/\xi)\, q_\mu q_\nu}{q^2 - m_W^2/\xi}\right]$$

$$\text{for G boson} \qquad \Delta^G(q;\xi) = \frac{1}{q^2 - m_W^2/\xi} \qquad (6.2)$$

$$\text{for H bosons} \qquad \Delta^H(q;\xi) = \frac{1}{q^2 - m_H^2}$$

Assuming $m_1, m_2, m_n, m_\lambda \ll m_W = m_H$ and neglecting the external momenta the final result for this crudest approximation is

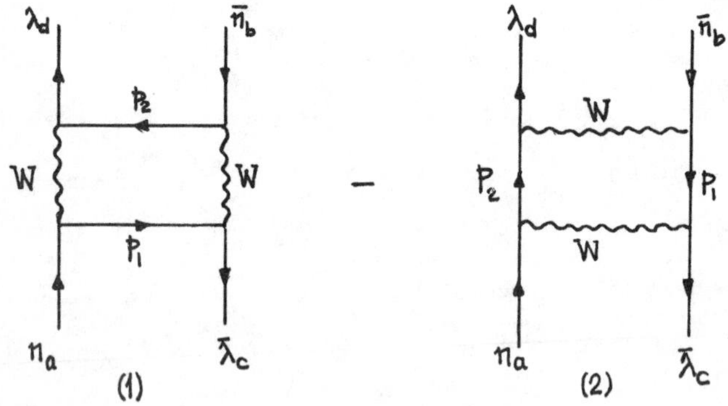

Fig.1. Diagrams for T conserving scattering of the $\Delta Y = \pm 2$ reaction: $K^o \rightleftarrows \bar{K}^o$. Each wavy line can be either W or G, each solid internal line can be either p_1 or p_2. Thus each diagram refers to 16 different assignments.

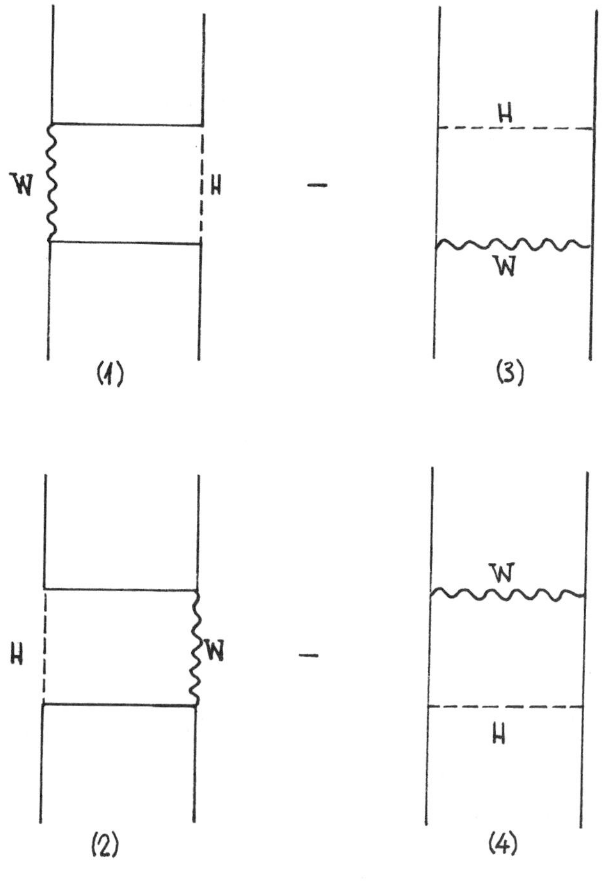

Fig.2. Diagrams for T violating scattering. Each wavy line can be either W or G, each solid internal line can be either p_1 or p_2. Thus each diagram refers to 8 different assignments.

$$M^{(+)}_{K\bar{K}} = \frac{iG_F^2 \cos^2\theta \sin^2\theta}{(2\pi)^2 (m_1^2 - m_2^2)} [m_1^4 - m_2^4 - 2m_1^2 m_2^2 \ln m_1^2/m_2^2]$$

$$M^{(-)}_{K\bar{K}} = \frac{2iG_F^2 \cos^2\theta \sin^2\theta}{(2\pi)^2 m_W^2 (\rho_1/\rho_2)^2} \{[2\ln m_1^2 - 1)m_1^2 + 2(\ln m_2^2 - 1)m_2^2] \kappa_n \kappa_\lambda^* - |m_1\kappa_1 + m_2\kappa_2|^2 \ln m_W^2\} \quad (6.3)$$

$$\varepsilon = 2\left(\frac{\rho_2}{\rho_1}\right)^2 \frac{m_1^2 - m_2^2}{m_W^2} \frac{[(2\ln m_1^2 - 1)m_1^2 + (2\ln m_2^2 - 1)m_2^2]\kappa_n \kappa_\lambda^* - |m_1\kappa_1 + m_2\kappa_2|^2 \ln m_W^2}{m_1^4 - m_2^4 - 2m_1^2 m_2^2 \ln m_1^2/m_2^2} \quad (6.4)$$

We observe that the imaginary part of ε comes from $\kappa_k \kappa_\lambda^*$ term, while $\ln m_W^2$ makes an important contribution to the real part. In order to have the phase equal to $\pi/4$, real and imaginary parts must have the same order of magnitude. That will give a strong constraint on the coupling constants k_2^h. Once this constraint is satisfied, the absolute value of ε comes out very approximately as

$$|\varepsilon| = 2\sqrt{2} \frac{\ln m_W^2}{m_W^2} \frac{(m_1^2 + m_2^2)^2}{(m_1^2 - m_2^2)} \quad (6.5)$$

ACKNOWLEDGEMENT

I wish to thank B. Stech, Y. Achiman and D. Gromes for helpful discussions.

REFERENCES

Abers, E.S. and Lee, B.W. [1973] *Physics Reports* 9,1.

Achiman, Y. [1973] *Nuclear Physics* B56, 635.

Glashow, S.L., Illiopoulos, J. and Maiani, L. [1970] *Phys. Rev.* D2, 1285.

Jacob, R. and Sachs, R.G. [1961] *Phys. Rev.* 121, 350.

Lee, T.D. [1973] *Phys. Rev.*, D8, 1226.

Lee, T.D. [1974] *Physics Reports* 9C, 143.

Sachs, R.G. [1965] *High Energy Physics and Elementary Particles*, Trieste Seminar, IAEA, Vienna, 929.

INDEX OF SUBJECTS